The Open University

Science: a third level course

GW01237625

S343 INORGANIC CHEMISTRY

Block 8 SOLID-STATE CHEMISTRY

The Open University

Prepared by an Open University Course Team

S343 Course Team

Course team chairs		*Stuart Bennett, Charlie Harding, Elaine Moore, Lesley Smart*
Authors	Block 1	*David Johnson*
	Block 2	*Elaine Moore*
	Block 3	*Kiki Warr, with contributions from David Johnson*
	Block 4	*Stuart Bennett*
	Block 5	*Michael Mortimer*
	Block 6	*Ivan Parkin, with contributions from Dr M. Kilner, Professor K. Wade* (University of Durham) *and F. R. Hartley* (Royal Military College of Science, Shrivenham)
	Block 7	*Paul Walton* (University of York), *with contributions from Lesley Smart*
	Block 8	*Lesley Smart, with contributions from David Johnson, Kiki Warr and Elaine Moore*
	Block 9	*David Johnson*
Consultants		*Dr P. Baker* (University College of North Wales)
		Dr R. Murray (Trent Polytechnic)
Course managers		*Peter Fearnley*
		Wendy Selina
		Charlotte Sweeney
Editors		*Ian Nuttall*
		David Tillotson
BBC		*Andrew Crilly*
		David Jackson
		Jack Koumi
		Michael Peet
Graphic artists		*Steve Best*
		Janis Gilbert
		Andrew Whitehead
Graphic designers		*Josephine Cotter*
Assistance was also received from the following people:		*George Loveday* (Staff Tutor)
		Joan Mason
		Jane Nelson (Staff Tutor)
Course assessor		*Professor J. F. Nixon* (University of Sussex)

The Open University, Walton Hall, Milton Keynes, MK7 6AA

First published 1989. Reprinted 1992, 2000

Edited, designed and typeset by the Open University.

Printed in the United Kingdom by Henry Ling Limited, at the Dorset Press, Dorchester, DT1 1HD.

ISBN 07492 5003 8

This text forms part of an Open University Level 3 course. If you would like a copy of *Studying with the Open University*, please write to the Course Reservations and Sales Centre, PO Box 724, The Open University, Walton Hall, Milton Keynes, MK7 6ZS, United Kingdom. If you have not enrolled on the Course and would like to buy this or other Open University material, please write to Open University Worldwide, The Berrill Building, Walton Hall, Milton Keynes, MK7 6AA, United Kingdom: tel. +44 (0)1908 858585, fax +44 (0)1908 858787, e-mail ouwenq@open.ac.uk. Alternatively, much useful course information can be obtained from the Open University's website http://www.open.ac.uk.

1.3

S343Block8i1.3

STUDY GUIDE FOR BLOCK 8

This Block has three components, the main text, articles from the S343 *Offprints Folder* and two video cassette sequences. The Block is the equivalent of two Units' or two weeks' work. It is concerned with the structure, properties and applications of solid materials, together with the relevant theory. Section 1 briefly revises some of the crystal structures studied in a Second Level Course, followed by a discussion of free-electron theory in Sections 2 and 3. Section 4 develops band theory. Sections 5 and 6 look in detail at how the presence of defects in solids can affect their properties. You should aim to finish Section 5 and study as much of Section 6 as possible by the end of the first week. The later material concerns magnetism (Section 7), one-dimensional conductors, superconductivity and zeolites (Section 8).

In addition to the main text, there are programmes relating to polyacetylene and to superconductors on Videocassette 2; zeolite chemistry is covered in a short sequence on Videocassette 1. Fuller details of the videocassette contents are to be found in the S343 *Audiovision Booklet*. The main text tells you when to watch the relevant videocassette sequence, but all relate to Section 8.

There are also a number of offprints in the S343 *Offprints Folder* providing up-to-date material on superconductors. You will be referred to these at the appropriate place in the text—again in Section 8.

You may find it helpful to use your model kit to build some of the structures discussed. Also, a spectrum is shown in colour in S343 *Colour Sheet 1*, and some of the structures in S343 *Colour Sheet 2*.

6

1 INTRODUCTION

Solid-state chemistry is concerned with the synthesis, structure, properties and applications of solid materials. Most of the chemical elements and their compounds are solids at room temperature. The study of solid materials is thus an extremely important area. The materials are usually inorganic, but not always: in this Block, we have included organic solids where they have particularly interesting physical properties. Currently, there is much research being pursued in the area of solid-state chemistry, partly because of the importance to industry of producing new materials: one area attracting much attention and excitement at the moment is that of high-temperature superconductors. We could not attempt to try and cover all the various properties and possibilities generated by solid-state research, and so we have chosen a few topics where we can relate physical properties such as magnetism and conductivity to the structure of the solid and in particular to the electronic structure of the solid. You will find that conductivity arises in two distinct ways. First, there is *electronic conductivity* which we met in a Second Level Course: here the current is carried by electrons that can move freely by virtue of the *band structure* of the solid. However, there is another mechanism by which current can be transmitted through a solid: this is via the movement of ions through the lattice, and this is called *ionic conductivity*.

SLC 1

Most solids form crystals where the atoms or molecules are packed in regular arrays in three dimensions. Understanding the structure and packing within crystals is central to this topic: it was dealt with in some detail in a Second Level Course, and in the next Section we shall spend time revising these topics. At Second Level, we also met the idea that crystals are not the ideal three-dimensional arrays that they are usually depicted to be. Indeed, crystals contain various types of imperfections in the lattice, collectively known as defects. The defects control, to a large extent, the properties of solids: they influence, in particular, electrical conductivity, chemical reactivity and mechanical strength. We consider in some detail the formation of point defects and their relationship to ionic conductivity in solids.

SLC 2
SLC 3

Many physical methods are used to explore the structure and properties of solids. There is not space in this Block to discuss the theory of any of the techniques, such as X-ray and neutron diffraction or electron microscopy. Many of the techniques used are either covered in other chemistry courses or are considered earlier in this Course. We confine ourselves to using the results.

The study of the structure and related properties of solids is an extremely fruitful area with many possibilities for the development of new materials. It is a truly interdisciplinary subject, where physicists, chemists, geologists, engineers and biologists all contribute to different aspects. In such a short space, we can do no more here than try to give you a flavour of the sheer breadth of the subject area and a sense of the excitement felt by those researchers who are discovering new materials and/or new properties that could have far-reaching technological implications.

To begin with, we must ensure that you have some of the 'tools of the trade' at your fingertips and can remember the important features of crystal lattices and how to interpret their crystal packing diagrams, so the next Section revises these topics.

1.1 Close-packing and related structures (*revision*)

A *close-packed layer* of spheres (representing atoms) is shown in Figure 1. In Figure 2, we have started to build up a structure by placing another close-packed layer on top. Notice that the new layer only covers half of the spaces in the first layer—those marked with a cross.

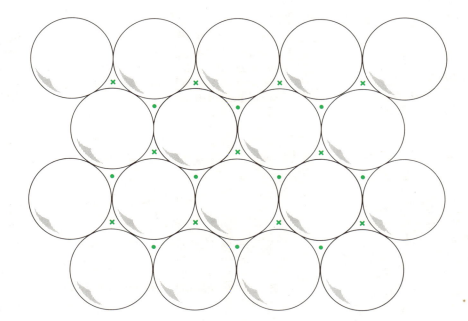

Figure 1 Close-packed layer of spheres.

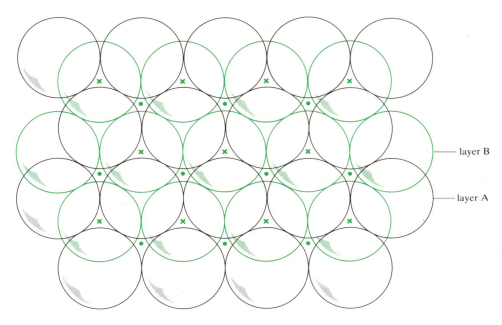

layer B

layer A

Figure 2 Two layers of close-packed spheres.

☐ When a third close-packed layer is added, where can it go?

■ There are now two alternative stacking sequences. In the first possibility, the third layer is directly above the first, giving a stacking sequence ABABAB . . . . This is known as *hexagonal close-packing, h.c.p.* The alternative is for the third layer to be positioned over those spaces marked with a dot. This gives a stacking sequence ABCABCA . . . and is known as *cubic close-packing, c.c.p.*

☐ Identify the *octahedral* and *tetrahedral holes* in Figure 2. How many of each are there?

■ The tetrahedral holes are marked with a cross, and there are $2n$ in an array of n atoms. The octahedral holes are marked with a dot, and in an array of n atoms there would be n.

☐ What is the *coordination number* of any atom in a close-packed structure?

■ 12.   Use Figure 2 to satisfy yourself that this is true.

Many metals adopt a close-packed structure. However, those that don't are usually *body-centred cubic* and (rarely) *primitive cubic*.

Crystal structures are usually depicted in one of two ways—either by a perspective drawing of the unit cell (a clinographic projection) or by a *packing diagram.* (A packing diagram is a two-dimensional projection of the contents of the unit cell, which is usually made by looking down one of the crystallographic axes.) This is demonstrated in Figure 3 for a body-centred (I) cubic unit cell.

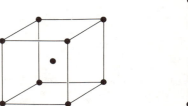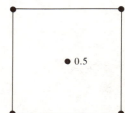

Figure 3 A body-centred (I) cubic unit cell and its packing diagram.

☐ Now draw similar diagrams for the other types of unit cells—primitive (P), face-centred (F), and face-centred (A or B or C).

■ The diagrams are shown in Figure 4.

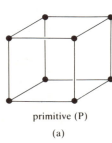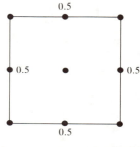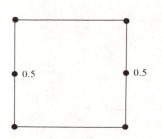

primitive (P)
(a)

face-centred (F)
(b)

face-centred
A (B or C)
(c)

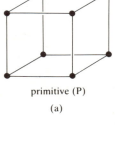

Figure 4 (a) Primitive, (b) face-centred, and (c) A-face-centred unit cells, together with their packing diagrams.

Many crystal structures of simple compounds are based on a close-packed array of one type of atom, with another (different) atom filling some or all of the tetrahedral and octahedral holes. The *sodium chloride (or rock-salt) structure* is the most obvious example, where Na^+ ions fill *all* the octahedral holes in a c.c.p. array of Cl^- ions. Table 1 lists some of the compounds that adopt the NaCl structure (of which there are more than two hundred in all).

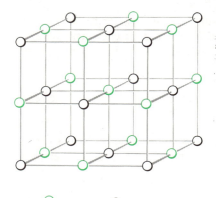

Na^+ (or Cl^-) Cl^- (or Na^+)

Figure 5 The crystal structure of sodium chloride, NaCl.

Table 1 Compounds that have the NaCl (rock-salt) type of crystal structure

all the alkali halides, MX, and AgF, AgCl, AgBr
all the alkali hydrides, MH
monoxides, MO, of Mg, Ca, Sr, Ba
monosulphides, MS, of Mg, Ca, Sr, Ba

☐ What kind of centring does the NaCl unit cell show? (Figure 5.)

■ F—face-centring with all faces centred.

SLC 4 In a Second Level Course, we used the idea of linking [SiO$_4$] *tetrahedra* together to form the many complex structures of the silicates. It is possible to view many of the structures in this Block as linked *octahedra*: each octahedron consists of a metal atom surrounded by six other atoms situated at the corners of an octahedron (Figure 6a and b). These are often drawn in diagrams, viewed from above with contours marked as in Figure 6c. Octahedra can link together via corners,

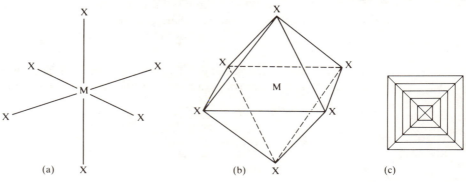

Figure 6 (a) An [MX$_6$] octahedron; (b) the solid octahedron; (c) plan of an octahedron with contours.

(a)   (b)   (c)

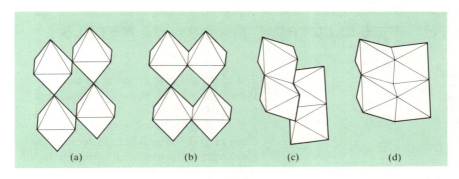

Figure 7 The conversion of (a) corner-shared MX$_6$ octahedra to (b) edge-shared octahedra and (c) edge-shared octahedra to (d) face-shared octahedra.

(a)   (b)   (c)   (d)

edges and faces, as shown in Figure 7. If you find this difficult to visualise from the diagram, it is quite easy to make small octahedra from paper and see how they join together, Figure 8 gives you a template to copy and cut out and glue.

The NaCl structure can be described using linked octahedra. In it, the [NaCl$_6$] octahedra share edges.

☐   How many edges does an octahedon have?

◼   12.

All the edges are shared in NaCl—each edge is shared by two octahedra. This is illustrated in Figure 9, which shows an NaCl unit cell with three of the [NaCl$_6$] octahedra picked out in green: the common edges are picked out by a thick broken line. The maximum number of regular octahedra that can meet at a point is *six*, and one of the resulting tetrahedral spaces is shown in green shading. If you find it difficult to convince yourself of this from the diagram, you should try building six identical octahedra and fitting them together. How can we determine

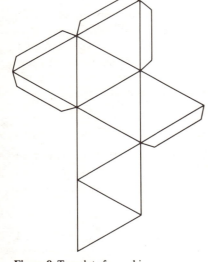

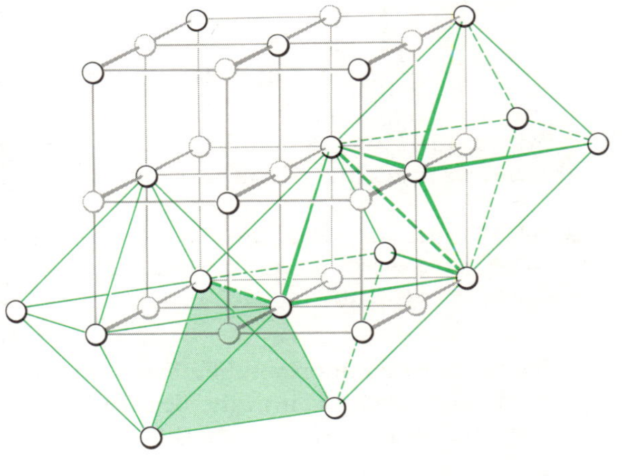

Figure 8 Template for making an octahedron.

Figure 9 Unit cell of the NaCl structure showing edge sharing of octahedra. (A tetrahedral space is also shown shaded green).

the overall stoichiometry of NaCl, from the linked octahedra? We know that a single Na^+ ion sits in the middle of each octahedron, coordinated by six Cl^- ions.

☐   How many Na^+ ions share each Cl^-?

■   Because all the edges are shared in NaCl, six octahedra meet at each vertex where there is a Cl^- ion. Thus each Cl^- ion is coordinated to six Na^+ ions, giving us the overall stoichiometry 1 : 1.

Attempt SAQ 1 now, which revises the detailed crystal structure of NaCl, before moving on to look at the electronic structure of solids in Section 2.

SAQ 1   (a) Draw a packing diagram for the NaCl unit cell shown in Figure 5. (b) Where do the close-packed layers of Cl^- ions lie? (c) How many formula units of NaCl are contained in the unit cell? (d) Describe the geometry around both Na^+ and Cl^- ions.

2 THE FREE-ELECTRON THEORY OF METALS

SFC 1

The electrical properties of metals can be explored by combining two strands of thought from the Science Foundation Course. There, metals were described as collections of positive ions immersed in a pool of electrons. These electrons act as a sort of glue, because it is the electrostatic attraction between their negative charges and the positive charges of the ions that holds the metal together. They also explain the electrical conductivity of the material: because the electrons can move freely throughout the metallic crystal, an overall drift of electrons is set up when a voltage is applied across two points on the metal surface.

This model of a metal is, of course, a rather crude one, but more valuable insights can be obtained by making it even cruder! The ion cores with their positive charges are regularly distributed throughout the crystal, perhaps, for example, in a cubic close-packed arrangement. Now, in toto, these positive charges are equal to the negative charges of the electrons. Let us therefore ignore the ion cores except in one respect: we allow their charges to cancel those on the electrons. Our model of a metallic crystal now consists simply of a set of decharged electrons moving freely in a container bounded by the crystal surface—we completely ignore the charges of the electrons, and the repulsive interactions between those charges. The advantage of this very crude model is that it gives us an opportunity to apply the quantum mechanics of a particle in a three-dimensional box.

2.1   Energy levels of particles in a box

SFC 2

In the Science Foundation Course, the hydrogen atom was treated in a way that is quite consistent with our model of a metal. It was regarded as a decharged electron confined within a cubic box of side L. The energy of the particle was entirely kinetic, and the allowed energy values were given as

$$E = \frac{h^2}{8mL^2}(n_1^2 + n_2^2 + n_3^2)$$

1

where h is Planck's constant, and m is the mass of the electron. The quantum numbers n_1, n_2 and n_3 can take only positive integral values 1, 2, 3, ... etc.

SAQ 2   (revision) What are the two lowest energy values for the particle in the box? What is the energy difference between these two lowest levels? (Express your answers as multiples of $h^2/8mL^2$.)

There are two important differences between our model of a metal and the model of the hydrogen atom in the Science Foundation Course. The first is the size of the box. A hydrogen atom is tiny, and in the Foundation Course, its single electron was confined within a cube with sides of 3×10^{-10} m. With this value of L, the energy separation of the two lowest levels calculated in SAQ 2 turns out to be 2.01×10^{-18} J or 12.5 eV. The experimental value obtained from the atomic spectrum of the real hydrogen atom is 10.2 eV. By contrast, a typical sample of a metal will be much larger. It could, for example, consist of a cube of side 3 cm $(3 \times 10^{-2}$ m). The electrons are now much less confined than in the hydrogen atom, and, according to our model, they are free to roam within a hollow cube of side 3×10^{-2} m.

SAQ 3 What is the energy difference between the two lowest energy levels for particles with the mass of an electron, confined within a hollow cube of side 3×10^{-2} m?

Thus, the energy levels of the conduction electrons for our model of a metal are very much more closely spaced than in the equivalent model of the hydrogen atom. This agrees with the principle introduced in the Foundation Course: in the metal, the conduction electrons are much less confined.

The second difference from the model of the hydrogen atom lies in the number of electrons. This is now very much larger. Consider, for example, metallic sodium. Its density is 0.971 g cm^{-3} and the relative atomic mass is 23.0. For our cube of side 3 cm, the volume is 27 cm^3 so,

$$\text{Amount of sodium} = 0.971 \text{ g cm}^{-3} \times 27 \text{ cm}^3$$

$$= 26.2 \text{ g}$$

This is 1.4 mol which, multiplying by the Avogadro constant $(6.022 \times 10^{23}$ mol$^{-1})$, is equivalent to 6.86×10^{23} atoms. Now in the Foundation Course, sodium was taken to be an array of Na$^+$ ions steeped in a pool of free electrons: there is one conduction electron for every sodium atom. Our box model of the chosen sample of metallic sodium therefore consists of a cube of side 3×10^{-2} m containing about 7×10^{23} electrons. This enormous number of electrons is distributed among the energy levels specified by equation 1. As with electrons that you have assigned to molecular orbitals, they obey the Pauli exclusion principle: they are allocated to the energy levels in pairs with opposed spin. The electron distribution can thus be obtained by putting two electrons into the lowest energy combination of n_1, n_2 and n_3, then two electrons into the (n_1, n_2, n_3) combination of next highest energy, and so on until all the electrons have been placed. In Figure 10, this has been done for 14 electrons which exactly fill the three lowest levels of energy. Alongside each level is given the sets of quantum numbers of the electrons allocated to it.

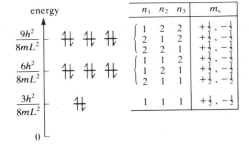

Figure 10 The electron energy states for the fourteen electrons of lowest energy according to the free-electron theory of metals. The table shows how each state is specified by a particular set of values of the quantum numbers n_1, n_2, n_3, combined with one of the two senses of spin designated $m_s = +\frac{1}{2}$ or $m_s = -\frac{1}{2}$.

2.2 The density of states

In Figure 10, only 14 of the 7×10^{23} electrons have been allocated to energy levels. Nevertheless, the Figure illustrates important ideas. First there is the phenomenon of degeneracy—of electron levels that have different values of n_1, n_2 and n_3 but the same energy. Thus the set of integers 2, 1, 1 can be allocated in three different ways to the quantum numbers (n_1, n_2, n_3); writing them in brackets in the order n_1, n_2, n_3, these ways are (2, 1, 1), (1, 2, 1) and (1, 1, 2). From equation 1, these three sets of quantum numbers yield three degenerate energy levels, and as each can take two electrons, six possible electron states with the same energy are allowed by the original set of integers, 2, 1, 1. As Figure 10 shows, the trio of integers 2, 2, 1 behaves similarly, because again, two of the three integers are identical.

☐ How many electron states are generated by a set of three identical integers?

■ Two; there is only one possible combination of the three numbers, and this gives a single energy level which can take two electrons. See, for example, the 1, 1, 1 level in Figure 10.

SAQ 4 Only one other kind of trio of integers remains: one in which all three integers are different as in the set 3, 2, 1. How many possible electron states arise from this set? Are they all degenerate?

We can now examine the pattern in the number of energy states as the energy increases. With the size of the box fixed, from equation 1, the energy is proportional to $(n_1^2 + n_2^2 + n_3^2)$, so if we list levels in order of this quantity, we list them in order of energy as well. In Table 2, this has been done up to states with the energy $27h^2/8mL^2$.

Table 2 The number of electron energy states associated with particular values of $(n_1^2 + n_2^2 + n_3^2)$ at low values of the integers n_1, n_2 and n_3

Permed integers	$(n_1^2 + n_2^2 + n_3^2)$	Number of states
1, 1, 1	3	2
2, 1, 1	6	6
2, 2, 1	9	6
3, 1, 1	11	6
2, 2, 2	12	2
3, 2, 1	14	12
3, 2, 2	17	6
4, 1, 1	18	6
3, 3, 1	19	6
4, 2, 1	21	12
3, 3, 2	22	6
4, 2, 2	24	6
4, 3, 1	26	12
5, 1, 1; 3, 3, 3	27	8

There exists an important quantity, which is concerned with how closely grouped the energy states are. Consider a narrow band of energy of width $5h^2/8mL^2$ centered on an energy $10h^2/8mL^2$. In Table 2, this will include any energy states with $(n_1^2 + n_2^2 + n_3^2)$ values between 8 and 12 inclusive. As Table 2 shows, such states exist only at the values 9, 11 and 12, and they total $(6 + 6 + 2)$ or 14 in all. Suppose, now, the band of width $5h^2/8mL^2$ is centred on $20h^2/8mL^2$.

☐ How many states does it now include?

■ The band now includes states with $(n_1^2 + n_2^2 + n_3^2)$ values 18–22 inclusive. Such states exist at values 18, 19, 21 and 22, and total $(6 + 6 + 12 + 6)$ or 30 in all.

Thus, *in this case*, an increase in energy is associated with more densely packed energy states. By trying SAQ 5, see now if this is observed with a further energy jump.

SAQ 5 Suppose the energy band of width $5h^2/8mL^2$ is centred on an energy of $40h^2/8mL^2$. Work out the total of allowed electron energy states within this range and insert your answer into the blank space in Table 3. (Remember that more than one trio of integers may give rise to the same $(n_1^2 + n_2^2 + n_3^2)$ value; see, for example, the two entries against 27 in Table 2.)

Table 3 The number of electron energy states included by a narrow band of width $5h^2/8mL^2$, centred on an increasing level of energy

$(n_1^2 + n_2^2 + n_3^2)$ for band centre	Number of states, $N(E)$
20	30
40	
90	72
300	134
400	162
600	180
1 000	270
2 000	360

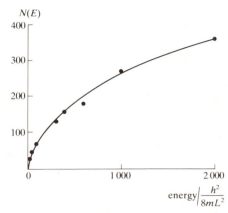

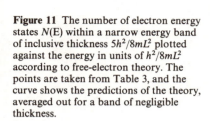

Figure 11 The number of electron energy states $N(E)$ within a narrow energy band of inclusive thickness $5h^2/8mL^2$ plotted against the energy in units of $h^2/8mL^2$ according to free-electron theory. The points are taken from Table 3, and the curve shows the predictions of the theory, averaged out for a band of negligible thickness.

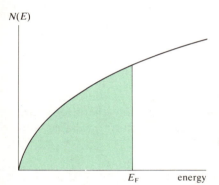

Figure 12 Use of the density of states curve to represent occupied electron energy levels in a metal at 0 K. The occupied levels are shaded green.

In Figure 11, the number of states within the narrow band is plotted against the energy on which the band is centred by using the data of Table 3. This curve is called the density of states function. It shows that the energy states do indeed become more closely spaced as the energy increases, although the change is much more marked at lower energies than at higher. Although we have only derived the curve for energies up to $2\,000h^2/8mL^2$, the general shape is correct for higher energies. We can now use this curve to represent the allocation of our 7×10^{23} electrons to the energy levels in a box of side 3×10^{-2} m. This has been done in Figure 12. We begin at the origin and progressively fill the electron energy states until all electrons have been allocated. With 7×10^{23} electrons, it turns out that the scale on the horizontal axis of Figure 12 runs from zero to about 10^{16}, compared with zero to $2\,000$ in Figure 11. The last energy states to be occupied have an energy denoted E_F, and this highest occupied energy level is called the **Fermi energy** or **Fermi level**. Strictly speaking, it is only at 0 K that all states below the Fermi level are occupied, and all states above it are empty. However, at normal temperatures, things are not so very different, so to simplify the argument we shall continue to assume a temperature of 0 K. Thus according to the free electron theory, the conduction electrons in our sodium sample have energies that cover a band of width E_F.

☐   Are there more electrons in the lower energy half or the upper energy half of this band?

■   The density of states curve in Figure 12 shows that there are more electrons in the upper half than in the lower.

2.3   Spectroscopy of conduction electrons

SFC 3
SLC 5

Let us now compare the deductions made using our very crude theory with some experimental results. In both the Science Foundation Course and at Second Level, the energy levels of electrons in compounds were investigated by photoelectron spectroscopy. The same method can be used to investigate energy levels in solids, but here it is more instructive to use the technique of **X-ray emission spectroscopy**.

In metallic sodium, the charge of the conduction electrons is balanced by that of the array of Na^+ ions which have the electron configuration $1s^2 2s^2 2p^6$. The electrons in, say, the 2p level of these ions are much more tightly bound than the outer conduction electrons, so they have a much lower energy. This point is illustrated in Figure 13, where the energy of the 2p level is placed well to the left of a density of states plot for the conduction electrons. In X-ray emission spectroscopy, metal samples are irradiated with a beam of electrons whose energy is sufficient to knock an electron out of the 2p level. The vacancy can be filled by a conduction electron. This is a transition to lower energy, so a photon is emitted. By determining the frequency, v, of the emitted radiation, one can obtain the energy of the transition from the relation

$$E = hv \qquad \qquad \textbf{2}$$

The energies are such that the emitted photons are X-rays.

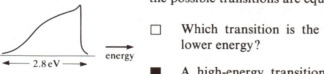

Figure 13 X-ray emission spectroscopy of sodium: a beam of high-energy electrons can create a vacancy in the 2p level, which is filled by a conduction electron. This transition to lower energy emits a photon in the X-ray region of the spectrum.

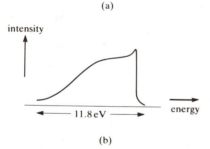

Figure 14 X-ray emission spectra obtained from (a) sodium metal and (b) aluminium metal when conduction electrons fall into the 2p level. The shapes are similar to those expected from free-electron theory.

Clearly, vacancies in the 2p level can be occupied through an electronic transition from any of the energy levels of the conduction band, so the X-ray emission spectrum of metallic sodium should consist of a band of width E_F. But not all of the possible transitions are equally probable.

☐   Which transition is the more probable; one of higher energy or one of lower energy?

■   A high-energy transition occurs from the upper part of the conduction band, where, according to the density of states function, there are more electrons. Higher energy transitions will therefore be more probable, and the intensity of their emitted radiation will be greater.

So, according to free-electron theory, the X-ray emission spectrum of a metal should contain a band whose intensity profile mirrors the density of states function in Figure 12. The experimentally observed bands for metallic sodium (a) and aluminium (b) are shown in Figure 14. Clearly, in these cases, the agreement with free-electron theory is very good.

SAQ 6 As noted in Section 2.1, the energies given by equation 1 are entirely kinetic. From the X-ray emission spectrum in Figure 14, the Fermi energy for aluminium is about 11.8 eV. Given that kinetic energy $E_k = \frac{1}{2}mv^2$ (where m is mass and v velocity) and taking any further information required from the S343 *Data Book*, what is the velocity of an electron in the Fermi level of aluminium?

2.4   Electronic conduction by metals

Consider the movement of electrons in a metal rod represented in Figure 15. SAQ 6 was a reminder that such a sample contains conduction electrons moving, in some cases, at very high speeds. At any instant, there will be electrons moving in all directions, some having a degree of motion to the right, and others a degree of motion to the left. However, *overall*, the degrees of motion to the right exactly balance those to the left, and so there is no *overall* movement of electrons along the rod in either of these two directions.

Figure 15 Electrons in a metal rod at zero electric field. They move in all directions, but overall there is no net motion to left or right: the *drift velocity* in these two directions is zero.

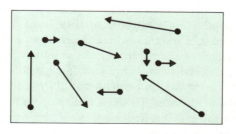

Suppose now that an electric field is set up in the rod by connecting the ends to the terminals of a battery (Figure 16). This constant electric field means that each electron now experiences a constant force in the left to right direction.

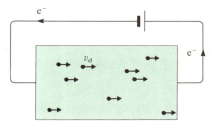

Figure 16 The sample of Figure 15 in a constant electric field, established by placing the rod between the terminals of a battery. The electrons continue to move in all directions, but now their velocities are modified so that each also has a net movement or drift velocity in the left to right direction.

☐ How will the electrons respond?

■ According to Newton's laws, they will accelerate to the right. Electrons with a degree of motion to the left will have their velocities steadily reduced; those with a degree of motion to the right will have their velocities steadily increased.

This implies that the conduction electrons will begin moving to the right with a gradually increasing **drift velocity**, v_d: the constant electric field produces a *progressively increasing* electric current. In fact, this behaviour is very rare. It is called superconductivity, and is observed in very special materials under special conditions. You will learn more about it later in the Block. What almost invariably happens instead is that the initial left to right acceleration of the conduction electrons is quickly curbed, and settles down into a *constant* left to right drift velocity: the constant electric field produces a *constant* electric current.

As you will see in a moment, to explain this curbing of the acceleration, we must invoke the *lattice of positive ion cores*. However, it is useful first to draw an analogy with the behaviour of a stone released in water. There is an initial acceleration, but the increasing velocity evokes an increasing *resistance* or friction from the surrounding water. This resisting force and the stone's weight quickly come into balance at a steady terminal velocity.

In the case of the electrons in a metal, the resistance is explained by saying that the electrons are scattered. This simply means that their direction of motion changes! To understand why, one must first accept that if the positive ion cores were arranged in a *perfectly regular* structure, the electrons could travel freely without scattering. This property is a consequence of the wave-like character of electrons. However, the ion cores do *not* have a perfectly regular structure: there will be impurity atoms present, the occasional defect such as a displaced ion core, and boundaries between the tiny crystals of which metal samples are usually composed. These irregularities cause scattering of the electrons, and in such cases one can think of this as the result of collisions with the impurities or defects.

SLC 6

At normal temperatures, however, the most important source of scattering is the coordinated vibration of the ion cores about their equilibrium positions. Such **lattice vibrations** also represent a departure from a perfectly regular structure; moreover, like the vibrations of molecules, which you met at Second Level, their vibrational energy is quantised; such a vibrational lattice mode is called a **phonon**. Consequently, a conduction electron can exchange energy with the vibrating system, changing direction at the same time. For example, the vibrating system can absorb a quantum of vibrational energy from an electron whose speed has been increased by the electric field. The electron's velocity decreases and changes direction; the amplitude of the vibration is increased. This increased amplitude and energy of vibration amounts to a rise in temperature of the metal. It is known as electrical heating or **Joule heating**, and is what happens whenever you switch on an electric fire.

Notice that this exchange of energy between the conduction electrons and the lattice vibrations also explains why the resistance of metals increases with temperature. At higher temperatures, the thermal energy causes the lattice to vibrate with a larger amplitude. These higher amplitude vibrations increase the departure from a perfectly regular structure. Both scattering and, therefore, resistance are then enhanced. At very low temperatures, the scattering caused by lattice vibrations is very much less. The resistance is then much smaller, and due mainly to scattering from impurities or defects.

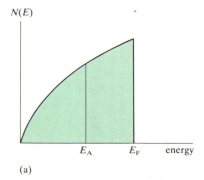

(a)

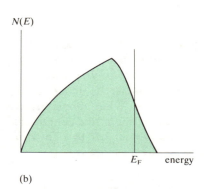

(b)

Figure 17 (a) The density of states diagram for a metal at low temperatures; an electron of energy E_A cannot contribute to electrical conduction because the states of slightly higher energy are full. (b) The modification of (a) in an electric field: the energy distribution of electrons near the Fermi level changes, as higher-energy states are occupied during conduction.

Finally, consider what electrical conduction implies for energy-level diagrams like Figure 12. You have seen, in Figure 16, that the initial effect of the electric field is to produce a steady increase in the velocity, and therefore in the energy of those conduction electrons with a degree of motion to the right. Now look at Figure 17.

☐ Why is such a steady increase impossible for an electron in the energy level marked E_A?

■ The steady increase requires the electrons to move to a *slightly* higher energy level, but the energy states immediately above E_A are completely filled by other electrons.

☐ So which energy levels contain the electrons that must carry the electric current?

■ The electrons at or very near the Fermi level; only these have access to vacant energy states of slightly higher energy.

So, during electronic conduction, there is a very slight initial redistribution of electron energies at or very close to the Fermi level. In terms of Figure 16, electrons with a degree of motion to the right move to slightly higher energy levels; those with a degree of motion to the left move to slightly lower energy levels. Scattering quickly stabilises the new distribution, and the overall result is to transform Figure 17a into Figure 17b. During electrical conduction, only electrons at or very near the Fermi level carry an electric current.

2.5 Summary of Section 2

1 In the free-electron theory of metals, a selected proportion of the electrons (the 'conduction' electrons) are considered to be 'decharged' and then assumed to move quite freely throughout the metal sample.

2 The particle in a box model gives the energy levels for such a system:

$$E = \frac{h^2}{8mL^2}(n_1^2 + n_2^2 + n_3^2)$$

1

where n_1, n_2 and n_3 take positive integral values. The conduction electrons are assigned to these levels using Pauli's exclusion principle. This means that each level specified by equation 1 takes two electrons with opposite spin.

3 When the conduction electrons are allocated to such states in order of increasing energy, successive energy states become more closely grouped. The energy at which all electrons have been allocated at 0 K is called the Fermi energy or Fermi level, E_F.

4 Electrical conduction by metals usually involves a constant electron drift velocity, v_d in a fixed electric field. The current is carried only by conduction electrons at or near the Fermi level.

3 INADEQUACIES OF FREE-ELECTRON THEORY

Free-electron theory provides valuable insights into the behaviour of conduction electrons in metals. However, the theory presupposes that the conduction electrons, which are 'free', can be sharply distinguished from the remaining electrons, which are not. With metals like sodium and aluminium, the prominent oxidation states of $+1$ and $+3$ guide us when we make this distinction. They suggest that there are one and three conduction electrons per metal atom, respectively. This is one reason why, as Figure 14 shows, the Fermi energy in aluminium is greater than that of sodium: because three times as many electrons must be allocated to the density of states diagram, the occupied energy levels span a larger energy range. With transition metals, however, there are signs that the behaviour of the conduction electrons is very different from what free-electron theory predicts. Thus, the configuration of the manganese atom is $[Ar]3d^54s^2$, which, when allied with the highest oxidation state of $+7$, suggests that in manganese metal there might be seven conduction electrons per atom. One might therefore expect the Fermi energy to be much greater than the 11.8 eV observed for aluminium. In fact, the experimental value obtained by X-ray emission spectroscopy is only 6 eV.

This difficulty in squaring free-electron theory with the properties of transition metals is quite general. In these metals, the motion of the conduction electrons is more constrained by the array of positive ion cores, and the conduction electrons spend much more time close to the ion core sites than free-electron theory implies.

4 BAND THEORY

With free-electron theory of metals, one starts with free electrons, and, when necessary, acknowledges constraint. Alternatively, one can start with constrained electrons, and then introduce freedom. You met this latter approach at Second Level. One begins with metal atoms containing electrons that are confined to atomic orbitals centred on the nuclei. From these atomic orbitals, molecular orbitals that extend over the entire metallic structure are formed. Thus the single outer electron in the sodium atom is in a 3s atomic orbital. When these 3s orbitals overlap in a crystal of sodium, they form a collection of bonding and antibonding molecular orbitals whose energies cover a band (Figure 18). At

SLC 1 Second Level, this theory was called *band theory*.

☐   If there are n atoms of sodium in the crystal, how many molecular orbitals are there in the band?

■ n atoms of sodium means n 3s atomic orbitals from which n molecular orbitals will be formed.

☐   How many of these n molecular orbitals are occupied by electrons in the sodium crystal?

■ $\frac{1}{2}n$; the band is half full. Each sodium atom contains one 3s electron, so the band contains n electrons. But each molecular orbital can take two electrons, so $\frac{1}{2}n$ molecular orbitals are occupied.

At Second Level, you learnt that the electrical conductivity of metals was due to *partly filled* bands. Notice how this agrees with the arguments of free-electron theory in Section 2.4: in both cases, the property is explained by the presence of many energy states immediately above the Fermi level. The chief difference is that, in free-electron theory, there is an ever-ascending stack of energy states above the Fermi energy; in band theory, the Fermi level of a metal lies within a band that has a finite upper energy limit.

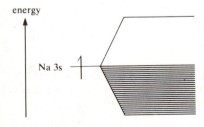

Figure 18 The conduction band in sodium metal; the energy of the 3s orbital in sodium atoms is shown left. In sodium metal, these orbitals overlap and generate a band of molecular orbitals, which is half full.

Unlike free-electron theory, band theory supplies a ready explanation for the existence of metals, semiconductors and insulators. Consider sodium chloride, an insulator, which is normally formulated as an ionic compound Na^+Cl^-. How is this related to a treatment based on band theory? We adopt the approach used at Second Level, and start with atomic orbitals for sodium and chlorine. The sodium atom has the configuration $[Ar]3s^1$, and the chlorine atom, $[Ar]3s^23p^5$. We consider only the highest-energy orbitals on each atom, and, as at Second Level, equate the orbital energies to minus the ionisation energies. For sodium, the required orbital energy is (3s: -8.2×10^{-19} J or -5.1 eV); for chlorine, it is (3p: -20.8×10^{-19} J or -13.0 eV). Thus an energy-level diagram (Figure 19) shows the Cl 3p orbital lying well below the Na 3s orbital.

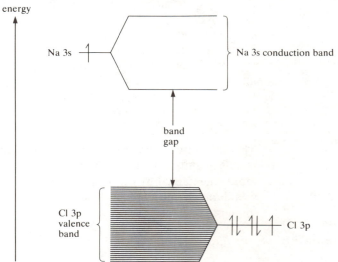

Figure 19 The band structure of sodium chloride developed through an ionic model.

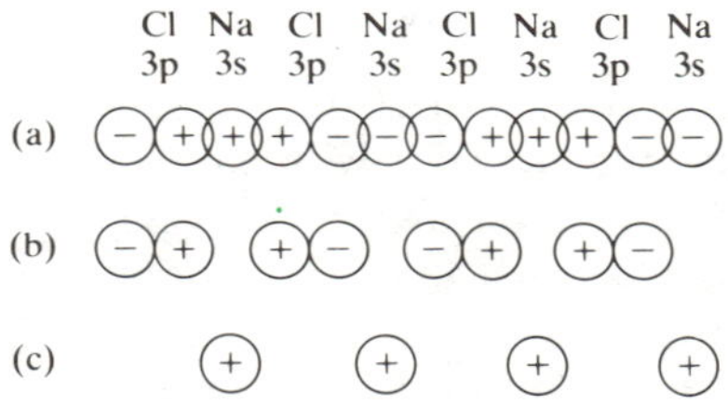

Figure 20 Possible bonding combinations of atomic orbitals in solid sodium chloride: (a) Na 3s with Cl 3p; (b) Cl 3p with each other; (c) Na 3s with each other. (The orbitals are depicted using angular functions.)

Now, the atomic orbitals are combined to form molecular orbitals for solid NaCl. One possibility is to combine Na 3s with Cl 3p; a bonding molecular orbital of this type is illustrated in Figure 20a for a chain of atoms in just one dimension. There is, however, a reason for playing down such combinations.

☐   Can you see what it is?

■   The Na 3s and Cl 3p orbitals differ markedly in energy.

There is a simple but crude approximation that avoids this problem by ignoring combinations of Na 3s and Cl 3p completely. It combines the Cl 3p orbitals just with each other, and then repeats the process for Na 3s. Figures 20b and 20c illustrate these procedures, which lead to the two bands shown in Figure 19; the lower band is formed entirely from Cl 3p; the upper from just Na 3s.

The final step is to assign electrons to the bands.

SAQ 7 If there are n NaCl formula units in the crystal, how many molecular orbitals will be associated with the lower band of Figure 19? State how many electrons this lower band contains, and explain why NaCl is an insulator.

Notice how the ionic model is expressed in the language of band theory. The electron in the 3s atomic orbital of sodium to the left of Figure 19 ends up in a band formed from just the Cl 3p orbitals: it is effectively transferred from sodium to chlorine.

The approximation that we used to obtain Figure 19 is too crude to be quantitative, and this is why no energy scale has been included. However, the photoelectron spectrum of NaCl, allied with theoretical calculations, suggests that the widths of the valence and conductance bands are about 4 eV and 6 eV respectively. The band gap, the photon energy required to excite an electron from the valence band to the conduction band, is about 8.5 eV.

Notice that the Na 3s conduction band is substantially broader than the Cl 3p valence band.

☐ Which of these two bands contains the greater number of electron energy states?

■ The valence band; with n NaCl formula units there are $6n$ states in the valence band (SAQ 7) but only $2n$ states in the conduction band.

Thus, the *density* of states is greater in the valence band on two counts: not only are there more states but they are compressed into a narrower range of energies. This point has been pictorially expressed in Figure 21, a density of states diagram, which also includes estimates of the band widths and band gap.

The greater width of the conduction band can be related to a tendency discussed earlier in the Course. Sodium occurs at the beginning of the third Period and chlorine at the end. As the nuclear charge increases across the row, orbitals tend to contract. Thus the Na 3s orbitals are more extended than Cl 3p, and can overlap with each other to a greater degree. At Second Level, you saw that greater overlap between atomic orbitals on two different atoms produces a greater separation of the bonding and antibonding molecular orbitals. Likewise, *in an extended solid, greater overlap between atomic orbitals on different atoms yields broader energy bands.*

SAQ 8 The gap between the valence and conduction bands in the sodium halides decreases in the order NaF > NaCl > NaBr > NaI, the actual values being 11.6 eV, 8.5 eV, 7.5 eV and 6.0 eV. Explain this decrease in terms of Figure 19. The visible region of the spectrum covers the energy range 1.8–3.1 eV. Explain why the four sodium halides are colourless.

4.1 Band theory of binary transition-metal compounds

Let us now extend the ideas of Section 4 to some transition-metal compounds. We begin with rutile, TiO_2, for which the extension is fairly straightforward. In this compound, titanium is octahedrally coordinated by oxygen (Figure 22a).

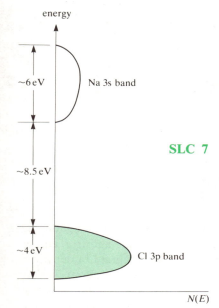

Figure 21 The band structure of sodium chloride in density of states form. (Note that whereas in Figure 12, energy is plotted horizontally and $N(E)$ vertically, here the opposite is the case.)

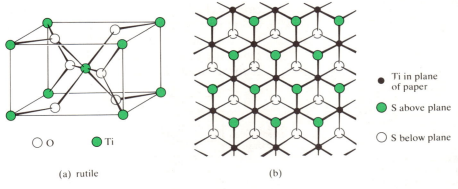

Figure 22 The structure of (a) rutile, and (b) the 3-deck layers in TiS_2 viewed from above.

A simplified band structure for TiO_2 is shown in Figure 23a (*overleaf*). As in the NaCl structure of Figure 19, the lowest band is formed from p orbitals on the electronegative element. Above come bands made up from valence orbitals of the metallic element. In titanium, these valence orbitals are 3d, 4s and 4p, but in this simplified treatment, we need consider only the 3d set, which lies lowest in energy. Now, in Figure 19, the 3s valence orbitals of sodium yielded a single broad band, but in Figure 23, you can see that *two* bands arise from the 3d orbitals of titanium.

☐ Can you suggest why this is so?

■ In TiO_2, the Ti 3d orbitals are in octahedral ligand fields. Just as they are split into t_{2g} and e_g levels by such a field in a discrete complex, so they are split into t_{2g} and e_g bands in a continuous solid.

Each TiO_2 unit supplies six O 2p orbitals for formation of the lower band, a total of twelve electron energy states per unit. Eight of these are filled by the 2p electrons of the two oxygens, and the remaining four by the four outer electrons

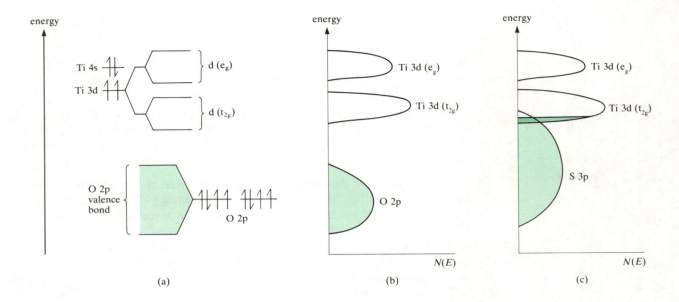

(a) (b) (c)

SLC 8

Figure 23 Band structures based on ionic models: (a) TiO_2; (b) as (a), but in density of states form; (c) one possible interpretation of the conductivity of TiS_2.

of titanium: once again, there is a filled valence band. The conduction band is the lower of the two bands formed principally from Ti 3d orbitals and it is empty: TiO_2 is an insulator. Alongside, in Figure 23b, is the same diagram in density of states form. Spectroscopic measurements suggest a band gap of about 3 eV or $25\,000\,cm^{-1}$. Plate 1.1 of S343 *Colour Sheet 1* shows that this is the high-energy limit of the visible spectrum, so TiO_2 is *just* colourless, a property exploited in its use as a pigment in white paint.

Now let us turn to TiS_2. This has the familiar CdI_2 structure, which consists of the three-deck layers of Figure 22b. It is useful at this point to reintroduce the description of the structure given at Second Level. This is done in Figure 24. One regards the sulphur atoms alone as hexagonally close-packed, and in Figure 24, each close-packed sulphur layer is represented by a horizontal line. Between these close-packed layers are octahedral holes, which, in total, are equal to the number of sulphur atoms. As Figure 24 shows, what happens is that a set of holes between one pair of sulphur layers is completely occupied, and the set between the next pair is left vacant. This pattern is repeated giving 50 per cent occupancy of the octahedral holes and the formula TiS_2.

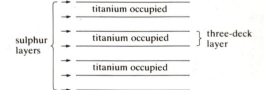

Figure 24 Occupation pattern of octahedral holes between close-packed layers of sulphur atoms in TiS_2.

For the moment, however, the important point to note about the structure is that, as in TiO_2, the titanium is in octahedral coordination, so the sequence of bands should be similar to that in Figure 23a. In addition, arguments of the kind used in SAQ 8 suggest that the band gap should be narrower in TiS_2, because the ionisation energy of sulphur is much less than that of oxygen.

☐ Is this prediction consistent with the fact that TiS_2 is golden-yellow?

◼ Yes; if there is a band gap, it must be less than 3 eV because absorption occurs in the visible region.

An even more striking fact is that the electrical conductivity of TiS_2 ($10^5\,\Omega^{-1}\,m^{-1}$) is greater than that of graphite. The explanation of this fact is controversial. One possible reason is that our predicted narrowing of the band gap goes so far that the valence and conduction bands *just* overlap. This possibility is pictured in Figure 23c: there are accessible vacant states immediately above the Fermi level, so TiS_2 is a metal or semimetal.

An alternative argument claims that the valence and conduction bands just *fail* to overlap, and attributes the good conductivity to a titanium content very slightly in excess of that prescribed by the formula TiS_2. Here, however, we shall promulgate the first explanation because it best fits the design of this Block!

4.2 The lithium–titanium disulphide battery

The electrical conductivity and other properties of TiS_2 have been exploited in attempts to design a new type of rechargeable electrical battery. As Figure 24 shows, in TiS_2, alternate layers of octahedral holes are occupied. The vacant layers of holes can accommodate small cations such as Li^+ in the reaction,

$$Li(s) + TiS_2(s) = LiTiS_2(s) \qquad\qquad 3$$

Reaction 3 can be carried out *electrochemically* in the cell shown in Figure 25. One electrode is made of metallic lithium, and the other of TiS_2 bonded together with a polymer such as Teflon. The electrolyte consists of a lithium salt such as $LiClO_4$ dissolved in an organic solvent, typically a mixture of dimethoxyethane (DME) and tetrahydrofuran (THF). When the circuit is completed, lithium metal dissolves to form solvated ions at the right-hand electrode, and lithium ions are deposited on the TiS_2 at the left:

$$Li(s) = Li^+(solv) + e^- \qquad\qquad 4$$

$$Li^+(solv) + TiS_2(s) + e^- = LiTiS_2(s) \qquad\qquad 5$$

$$\overline{Li(s) + TiS_2(s) = LiTiS_2(s)} \qquad\qquad 3$$

Notice the value of the high electrical conductivity of TiS_2: both electrode materials need to be good conductors.

The progressive insertion of lithium into the space between the sulphur layers in TiS_2 is called *intercalation*. The final product, $LiTiS_2$ can be represented by Figure 26 and is called an **intercalation compound**. Intercalation scarcely affects the original three-deck layers of TiS_2; the lithium injection merely causes a 10 per cent expansion of the unit cell dimension in the vertical direction in Figures 24 and 26.

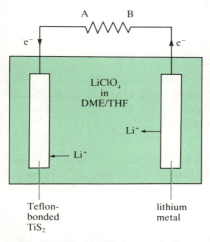

Figure 25 The Li–TiS_2 battery during a discharge phase.

Figure 26 Occupation pattern of octahedral holes between close-packed layers of sulphur atoms in the intercalation compound $LiTiS_2$.

sulphur layers	
	titanium occupied
	lithium occupied
	titanium occupied
	lithium occupied
	titanium occupied
	lithium occupied

A crucial property of the cell in Figure 25 is that its reaction can be reversed by replacing the resistor in the gap AB by an electrical power source that drives current through the system in the opposite direction. Then lithium is removed from $LiTiS_2$ and forms Li^+(solv) finally regenerating TiS_2. At the same time, solvated lithium ions are deposited on the metallic lithium: reactions 3, 4 and 5 are all reversed. This allows the battery to be recharged when its energy has been dissipated. The discharge/recharge cycle has been successfully repeated over 1 000 times on individual cells.

Because lithium is such a powerful reducing agent, the cell will deliver voltages comparable with that of the familiar, rechargeable lead–acid battery. But because both lithium and TiS_2 have relatively low densities, the same energy can be delivered from a cell of much lower mass. Understandably, therefore, research into cells of this type has been vigorous; however, in Section 5, you will see that they confront strong competition.

SAQ 9 If lithium is intercalated as Li^+, then $LiTiS_2$ can be written $Li^+[TiS_2]^-$. Assuming this is legitimate, use Figure 23c to predict whether the electrical conductivity of the TiS_2 electrode in Figure 25 will increase or decrease as the cell is discharged.

4.3 Transition-metal monoxides

In recent years, metallic bonding has been detected in simple compounds such as oxides which were once thought of in ionic or covalent terms. This is especially true of compounds of early transition elements such as titanium, vanadium, zirconium and molybdenum. A good illustration of this is provided by the lattice energies of the monoxides and dichlorides of the elements calcium to zinc. These lattice energies are values of $\Delta H_m^{\ominus}$ for the reactions:

$$M^{2+}(g) + O^{2-}(g) = MO(s) \qquad\qquad 6$$

$$M^{2+}(g) + 2Cl^-(g) = MCl_2(s) \qquad\qquad 7$$

In most of these oxides and chlorides, the metal ion is in octahedral coordination.

☐ How, then, would you expect the two lattice energies to vary between calcium and zinc?

■ In both reactions, the gaseous metal ion enters into octahedral coordination. Plots of both should show a double-bowl shape, the individual points being displaced below a curve through the Ca, Mn and Zn values by the octahedral ligand field stabilisation energies.

As Figure 27 shows, these predictions are largely correct. But there is one obvious anomaly: the monoxide bowl in the first half of the series is abnormally large, much larger than in the second. In other words, TiO and VO have more negative lattice energies than expected. The source of this additional stability is not hard to find. Our prediction of the lattice energy variation relied on a formulation $M^{2+}O^{2-}$, each M^{2+} ion having a configuration of the type $[Ar]3d^n$ in which the d electrons are localised on the metal-ion sites. But TiO is a black, lustrous solid with a high metallic conductivity.

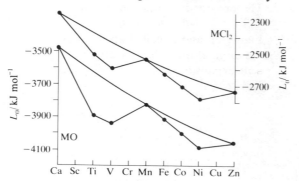

Figure 27 The lattice energies of the dichlorides of the elements Ca to Zn (upper plot; right-hand axis) and of the monoxides (lower plot; left-hand axis). The Jahn–Teller distorted compounds of chromium and copper have been omitted.

SAQ 10 The orbitals used to construct a band structure for TiO and TiO_2 will be the same*, so in general outlines, Figure 23a is applicable to both. Use it to explain why TiO is metallic, and specify the percentage occupancy of the occupied bands.

This interpretation of the bonding in TiO implies that the formulation $Ti^{2+}O^{2-}$ is inappropriate because the two d electrons of the $[Ar]3d^2$ configuration of the Ti^{2+} ion are not localised on transition-element sites. Instead, they are delocalised over the entire crystal in a band formed from the $d(t_{2g})$ orbitals. The lowering of the energy of occupied orbitals due to band formation is an additional stabilisation, which shows up in the lattice energy plot in Figure 27. (The explanation for VO, which is also metallic, is very similar.) To retain a 'simple' formulation to replace $Ti^{2+}O^{2-}$, we could distinguish the delocalised conduction electrons as in free-electron theory, and write the compound, $Ti^{4+}(2e^-)O^{2-}$.

The existence of metallic bonding in TiO is apparent not only in the compound's metallic conductivity, but also in its structure. Ignoring TiO and VO, the monoxides treated in Figure 27 have the NaCl structure (Figure 5). In TiO, however,

* One possible difference is that in the lower oxidation state, the splitting of localised $d(t_{2g})$ and $d(e_g)$ orbitals will be smaller, so that the bands formed from them may merge into a single 3d band. However, this does not affect the arguments used here.

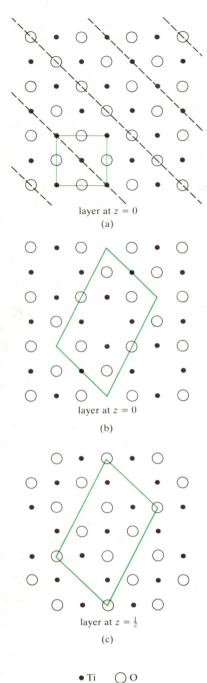

layer at $z = 0$

(a)

layer at $z = 0$

(b)

layer at $z = \frac{1}{2}$

(c)

● Ti   ○ O

Figure 28 Layers parallel to the horizontal planes of Figure 5. (a) The hypothetical TiO structure of the NaCl type shown in Figure 5; the line of intersection of every third vertical diagonal plane is marked by a broken green line. (b) The same plane in the observed structure of TiO; every alternate atom is removed along the diagonal lines in (a). (c) The plane directly beneath the layer in (b); again, every alternate atom is removed along the cuts made by the planes whose intersection lines are shown in (a). In (b) and (c), the cross-section of a monoclinic unit cell is indicated.

this structure is modified by removing equal numbers of titaniums and oxygens in an ordered fashion.

Figure 28a shows the top (or bottom) face of an NaCl unit cell extended in space with the face of just one unit cell picked out. Every third *vertical* diagonal plane is marked by a broken green line. Notice that the lines of intersection of these vertical planes with the horizontal alternately contain all titaniums and all oxygens. Figure 28b shows how this face looks in the observed structure of TiO: *every other atom* has been removed along both the diagonal titanium lines, and the diagonal oxygen lines.

Figure 28c shows how the *central* horizontal plane of the unit cell in Figure 5 is affected. Where the lines of intersection with the horizontal planes above and below contain just titaniums (or oxygens), here they contain just oxygens (or titaniums). Again, along these diagonals, every other atom is removed. Figure 28 therefore shows that the structure of TiO can be formed by removing one half of the titaniums and one half of the oxygens from every third diagonal plane.

SAQ 11   What *total* fraction of titaniums and oxygens must be removed from the hypothetical TiO structure of Figure 5 to generate the observed structure?

This ordered elimination of one-sixth of the atoms from an NaCl structure gives rise to a new unit cell whose top and bottom faces are the parallelogram picked out in Figure 28b. Its intersection with the intermediate plane is shown in Figure 28c. Such a unit cell is said to be *monoclinic*: two faces are parallelograms, and four are rectangles.

Finally, one must ask why it is that for TiO, the deficient NaCl structure is more stable than a fully occupied one. The most favoured explanation is that the removal of some atoms allows the structure to contract, reducing the internuclear distances between those atoms that remain. Thus, the overlap between the 3d orbitals on titanium is increased, broadening the conduction band and lowering the energy of its occupied levels.

SAQ 12 Ti_2O_3, WO_3 and ReO_3 all have structures in which the metallic element is octahedrally coordinated by oxygen. Predict whether they will be metals or insulators, and in each case, specify the percentage occupancy of the occupied bands.

SAQ 13   The black metallic oxide, NbO, also has a deficient NaCl structure, shown in Figure 29. By comparing it with Figure 5, calculate the fraction of niobiums and oxygens that have been removed from the fully occupied NaCl structure.

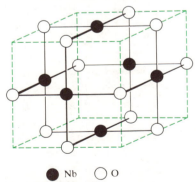

● Nb   ○ O

Figure 29 The unit cell of NbO. The broken outline shows the edges of an NaCl-type unit cell from which niobiums and oxygens have been removed.

4.4   Band width and orbital overlap

Broad bands are produced by a large overlap between the atomic orbitals from which the band is built up. The band structures used in Section 4 have been developed through an ionic model: a lower-energy valence band was formed solely from the valence orbitals of a non-metal like oxygen, and a higher-energy conduction band solely from those of a metallic element like titanium. However, in compounds such as ReO_3, the large proportion of oxygen in the structure keeps the rhenium atoms well apart. Consequently, substantial overlap of Re 5d orbitals, of the kind invoked in SAQ 12, seems unlikely. It is time for a little more sophistication.

In Section 4.1, you were told that TiS_2 has metallic levels of conductivity because valence and conduction orbitals just overlap. This occurs partly because the Ti 3d and S 3p orbitals are closer in energy than the Ti 3d and O 2p orbitals. There is, however, a second and more subtle reason. Because the Ti 3d and S 3p orbitals are closer in energy, the wavefunctions of the bands labelled Ti 3d will contain significant contributions from the S 3p orbitals. Likewise, the so-called S 3p band will contain a contribution from Ti 3d orbitals. This introduces an element of overlap between orbitals on *adjacent* atoms. It can therefore be thought of as a covalent interaction, which broadens the bands and makes their overlap more probable.

Band formation from orbitals on both the metallic and non-metallic atoms in a compound is more likely when the electronegativities of the two kinds of atom are not too different, and when an ion formulation of the compound gives ions of higher, and perhaps unrealistic charge (e.g. ReO_3, which would contain Re^{6+} ions, presumably has a conduction band built up from both Re 5d and O 2p orbitals).

The first condition suggests that sulphides, selenides and tellurides are more likely to be metallic than oxides. You have seen that TiO_2 is an insulator and that TiS_2 has semimetallic conductivity; it should therefore be no surprise to find that $TiSe_2$ and $TiTe_2$ are semimetals.

The second condition suggests that oxides and sulphides are more likely to be metallic than halides. A comparison of metallic VO with the insulators VF_2 and VCl_2 is an example of this. As you will see, in Section 4.6, some halides are metallic, although they are usually insulators with narrow bands.

4.5 Limitations of band theory

According to Section 4.3, a transition-metal monoxide, MO, will be metallic, whenever Figure 23a predicts the presence of partly filled bands. Unfortunately, such predictions are not always accurate.

☐ MnO has the NaCl structure; does Section 4.3 imply that it should be metallic?

■ Yes; two of the seven outer electrons per manganese help to fill the O 2p valence band of Figure 23a. The remaining five give only five-sixths occupany of the $d(t_{2g})$ band.

In fact, MnO is an insulator. This failure can be traced to a weakness that you have already encountered in a different context: orbital energies make an insufficient allowance for interelectronic repulsion. Figure 30 clarifies this by inserting two new steps into Figure 23a. As before, the valence band is formed directly from the O 2p orbitals, but band formation by the metal d orbitals is carried out in three stages, A and B and C. In A, the d electrons remain localised on the manganese sites, and the d orbital energies are merely split into the t_{2g} and e_g sets by the local octahedral ligand field of the six surrounding oxygens. The five electrons in these orbitals take up a high-spin configuration. This stage could also be reached by crystal-field theory in which the compound is written $Mn^{2+}O^{2-}$, the five d electrons are those of Mn^{2+}, and the high-spin configuration is adopted because O^{2-} is a weak-field ligand. In B, the d electrons localised on the manganese sites have been paired up *as far as possible*. The purpose of this step is to mirror, in a localised sense, the pairing of electrons that would ultimately occur if bands were formed from the metal d orbitals. In this particular case, the pairing yields a d electron configuration that is identical with that of octahedral low-spin d^5. Finally, in stage C, t_{2g} and e_g bands have been formed by d orbital overlap, and all the d electrons become delocalised in the t_{2g} band.

Figure 30 A simplified band structure for a hypothetical metallic form of MnO, formed in 3 stages, A, B and C. In all three cases, there is an O 2p valence band. In A, the d electrons remain isolated on individual manganese sites, the d orbitals are split by the ligand field, and a d^5 high-spin configuration is formed. In B, the d electrons, still localised, are paired up *as far as possible* in the t_{2g} orbitals. In C, the manganese d orbitals are allowed to overlap and form bands, so that the t_{2g} band is partly filled.

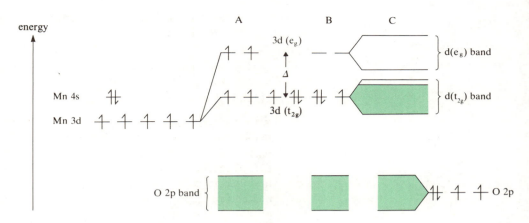

☐ In terms of orbital energies alone, does stage A or stage C have the lower energy?

■ Stage C; in the conversion of A to C, the two e_g electrons first pair up with electrons in the t_{2g} level as in B. This lowers the orbital energy by 2Δ per manganese. There is then, in stage C, a further lowering on $d(t_{2g})$ band formation because the centre of gravity of the t_{2g} band lies below the energy of the isolated t_{2g} orbitals.

But you know from earlier in the Course that orbital energies are not everything. In moving from A to B, there is an unfavourable change in interelectronic repulsion. In the language used earlier, the energy increases by $2P$ per manganese, where P is the repulsion energy required to pair each pair of electrons in the t_{2g} band.

In a discrete manganese(II) complex such as $[Mn(H_2O)_6]^{2+}$ where the ligands are weak-field ones, the high-spin state is preferred because the lower orbital energy of the low-spin state does not compensate for the increased interelectronic repulsion. In MnO, the metal sites are closer together, and band formation by metal d orbital overlap offers a further lowering of orbital energy. It seems, however, that this is still not enough. Consequently, stage A is a better description of MnO than stage C: sets of five 3d electrons remain localised with parallel spins on the manganese sites, and the compound is an insulator. Experimental evidence of this is provided by the magnetic moment, which, at high temperatures, is $5.9\mu_B$ in agreement with the formula, $\mu_s = [n(n+2)]^{1/2}\mu_B$, introduced earlier in the Course.

SAQ 14 Draw a version of Figure 30 for the case of the compound NiO, which has the NaCl structure. Explain why NiO is an insulator. What feature of your diagram would have to be increased to make NiO a metal?

As the case of NiO shows, band width and interelectronic repulsion are the competing factors that decide whether this type of transition-metal compound is to be an insulator or a metal. In this context, it is interesting to compare NiO with the metals TiO and VO. If TiO were an insulator, it would have two localised d electrons at the metal site like NiO; if NiO were a metal, it would have a half-filled band like VO.

☐ How would you expect the 3d band widths in TiO and VO to compare with that in NiO?

■ They must presumably be larger if they compensate for the interelectronic repulsion that makes NiO an insulator.

☐ Why should the band widths be larger at the beginning of a transition series than at the end?

■ Large band widths are due to large orbital overlap. Across the series, the 3d orbitals contract under the influence of increasing nuclear charge. Thus, band widths should be greater for TiO than for NiO, and metallic properties more likely.

This phenomenon of d-orbital contraction can, therefore, account for the fact that TiO and VO are metals, and that from manganese onwards, the monoxides of the first transition series are insulators or semiconductors.

We conclude this Section by mentioning a limitation in our explanation of the insulating properties of oxides such as NiO: in Figure 101, the interelectronic repulsion penalty incurred in forming a metallic state is taken to be a spin-pairing energy between e_g electrons at the same Ni^{2+} site. But the electronic mobility in a metallic form of NiO would give rise to charge fluctuations at nickel. For example, an e_g electron on one nickel site could move to an e_g orbital on another, momentarily altering its electronic configuration from d^8 to d^9. As Figure 31a shows, the transferred electron must again incur an unfavourable pairing energy of a kind not considered in Figure 101. This effect also contributes to the suppression of conductivity by interelectronic repulsion in NiO. Moreover, as you will see in Section 6, it is possible to make non-stochiometric NiO, which contains a small amount of nickel(III). There are then a few d^7 sites in the lattice, and when an electron is transferred to these from a nickel(II) site (Figure 31b), no pairing is required. Electron hopping is then easier, and it can be achieved by heating the oxide. Non-stoichiometric NiO is therefore a semiconductor.

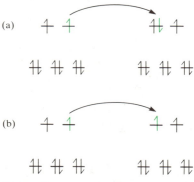

Figure 31 Electron transfer between two localised Ni^{2+} sites leads to electron pairing (a), but transfer from Ni^{2+} to Ni^{3+} sites does not (b).

4.6 Metal–metal bonding in compounds of the early transition elements

In the last Section, you saw that early in the first transition series, d-orbital extension led to metallic monoxides. Now, d-orbital extension will be greater in the second and third transition series, where the principal quantum number is greater. We might, therefore, expect metal–metal bonded compounds to be especially common for early second-row and third-row transition elements. This is the case: it occurs even among halides. Thus in the first transition series, copper, at the end of the series, is the only element to form monohalides, and the compounds are insulators. In the second-row and third-row transition series, monohalide insulators again occur at silver and gold, but in addition, zirconium and hafnium monohalides exist and they are metallic.

SLC 2

The structure of ZrCl is shown in Figure 32. To understand it, imagine a cubic close-packed array of spheres. At Second Level, you saw that the repeat pattern for cubic close-packing is ABCABCABC .... The structure of ZrCl can be obtained by substituting chlorine atoms for the spheres of layer 1, zirconium atoms for those of layers 2 and 3, and chlorine atoms for those of layer 4. This gives the four-deck layer of Figure 32, and the substitution pattern is then repeated through the cubic close-packed structure:

$$| ClZrZrCl | ClZrZrCl | ClZrZrCl | ....$$

☐ How do the chlorine atoms in Figure 32 reflect the ABC repeat of cubic close-packing?

■ In the ABC repeat, exact superposition occurs every 4th layer. In the | ClZrZrCl | layer of Figure 32, the halogens of layer 4 exactly cover those of layer 1.

SAQ 15 Layer 4 of the ZrCl structure is exactly superposed on layer 1 *and* contains the same type of atom. What is the number of the next layer for which this double matching occurs?

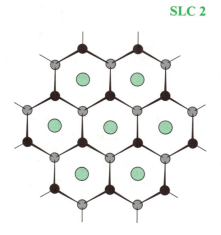

◯ Cl: layers 1 and 4

◯ Zr: layer 2

● Zr: layer 3

Figure 32 A plan view of the layers in the ZrCl structure. The centres of the atoms are similar in position to those of atoms in a cubic close-packed structure. In a sense, four layers are shown, because the chlorines of layer 4 are exactly superposed on those of layer 1.

The simplest possible approach to the band structure of ZrCl would follow that used for NaCl in Figure 19. There would be a full Cl 3p valence band, leaving three electrons per zirconium to partly fill a conduction band built up from Zr 4d orbitals: in free-electron notation, $Zr^{4+}(3e^-)Cl^-$. As the conduction electrons are in a Zr 4d band, the metallic conductivity should be confined to the zirconium

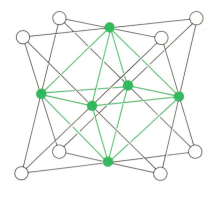

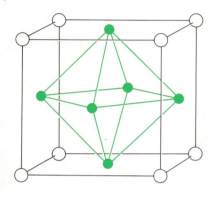

○ Cl   ● Mo

Figure 33 Two views of the $[Mo_6Cl_8]^{4+}$ cluster. The lower one emphasises the fact that the eight chlorines lie at the corners of a cube with the molybdenums approximately at the centres of the faces.

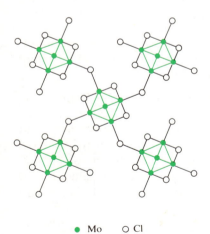

● Mo   ○ Cl

Figure 34 The linking of $[Mo_6Cl_8]^{4+}$ clusters in the layer structure of $MoCl_2$. Each cluster shares four chloride ions with four other clusters within its layer. There is also an unshared chloride above and below along the central vertical axis of each cluster (not shown).

layers. It should thus be *two dimensional*, occurring parallel to the plane of the paper in Figure 32, but not at right angles to it.

This is correct. The binding forces between |ClZrZrCl| layers are weak, so the crystals readily shear parallel to them, forming thin plates. The conductivity parallel to the plates is about $6 \times 10^3 \Omega^{-1} m^{-1}$, approximately 40 000 times the value perpendicular to the plates.

In some cases, the existence of metal–metal bonding in second-row and third-row transition-metal halides does not result in metallic conductivity. A comparison of $CrCl_2$, $MoCl_2$ and WCl_2 demonstrates this. $CrCl_2$ has a structure in which chromium is surrounded by a distorted octahedron of chlorines. $MoCl_2$, however, can be described by the formula $[Mo_6Cl_8]^{4+}(Cl^-)_4$, the $[Mo_6Cl_8]^{4+}$ unit being pictured in Figure 33. There is a central cluster of molybdenum atoms at the corners of an octahedron, and each molybdenum is also bound to four chlorines at the corners of a square. The Mo–Mo distance of 300 pm shows that strong metal–metal bonding exists, but because the $[Mo_6Cl_8]^{4+}$ units are separated from each other by sheaths of chloride ions (Figure 34), this metal–metal interaction is localised within the octahedral clusters and macroscopic metallic conductivity is not observed. WCl_2 has the $MoCl_2$ structure, and similar examples occur among the lower halides of niobium and tantalum.

Many of these metal–metal bonded systems have common structural features. Clusters of the type shown in Figure 33 are especially prominent, because not just the metallic but also the electronegative element can be varied. Thus, chlorine can be replaced by other halogens and by O, S, Se and Te.

The hollow octahedron of metal atoms is an even more common structural feature of metal–metal bonded systems of the early transition elements. Not only does it occur at the centre of the $[Mo_6Cl_8]^{4+}$ cluster (Figure 33); it also appears, joined by corners in the extended structure of NbO (Figure 29), and joined by edges in the doubled sheet of metal atoms in ZrCl, ZrBr and HfCl (Figure 32).

In conclusion, notice how in these metallic systems, the association of oxidation state with a particular isolated d^n configuration has collapsed. In ZrCl, for example, we do not have Zr^+ ions with $[Kr]4d^3$ configurations: the three d electrons per metal atom are paired in a band and delocalised throughout the material. Thus, these materials expose a limitation in something that has been a mainstay of many of our interpretations in earlier Blocks.

SAQ 16 When titanium and tellurium are strongly heated together in the correct proportions, a compound is formed that consists of chains in which clusters of the type in Figure 33, Ti_6Te_8, are linked through opposite corners of the metal octahedra. What is the formula of this compound?

4.7   Summary of Section 4

1   Electrons in solids are not free: their motion is affected by the attraction of the array of positive ion cores which breaks up the allowed energy states into bands.

2   Metallic conductivity is due to partly filled bands.

3   When a band structure of a binary compound of a metallic and an electronegative element is derived using an ionic model, the valence band is formed from the valence orbitals of the electronegative element and the conduction band from those of the metallic element.

4   If a partly filled band is narrow, this tends to keep the electrons localised around the ion cores. This yields insulators or semiconductors rather than metals.

5   Broad conduction bands and metal–metal bonding are more likely at the beginning of a transition series, and especially at the beginning of the 2nd and 3rd series where the d orbitals are more extended.

5 SOLID ELECTROLYTES

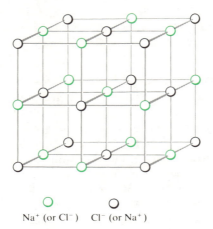

Na⁺ (or Cl⁻)   Cl⁻ (or Na⁺)

Figure 35 The crystal structure of NaCl.

5.1 Introduction: what is a solid electrolyte?

So far, we have concentrated on developing a picture of solid materials that goes a long way toward explaining their *electronic conductivity*—be they metallic conductors, or semiconductors, or insulators. But when a solid is ionic, there is another, and quite distinct, mechanism by which current can be transmitted through it.

A familiar ionic solid is sodium chloride (Figure 35), an array of Cl^- and Na^+ ions. Since ions are charged, they will respond to an electric field imposed across the crystal as do the electrons in a metal—although this time, of course, the negative ions will experience a net force in one direction, and the positive ions in the other. The resulting movement of ions that constitutes an electric current is just what happens when an ionic solid is dissolved in water (to give an electrolyte solution) or melted (to form a molten salt)—and then electrolysed.

Table 4 Typical values of electrical conductivity

	Material	Conductivity/$\Omega^{-1}\,m^{-1}$
Ionic conductors	ionic crystals	$<10^{-16}$–10^{-2}
	solid electrolytes	10^{-1}–10^{3}
	strong (liquid) electrolytes	10^{-1}–10^{3}
Electronic conductors	metals	10^{3}–10^{7}
	semiconductors	10^{-3}–10^{4}
	insulators	$<10^{-10}$

Until recently, the ionic conductivity of solids attracted little interest, because, as Table 4 shows, the conductivities of most ionic solids are orders of magnitude lower than the electronic conductivities typical of metals and semiconductors. A further contrast with the conductivity of metals is that **ionic conductivity** (symbol κ, Greek 'kappa'), can *always* be enhanced—usually greatly—by raising the temperature. Some typical experimental results—for crystalline NaCl in this case—are reproduced in Figure 36. The reasons for presenting the data in this way will become clearer later on. For the moment, note only that the temperature increases from right to left—and that a rise of some 400 K pushes up the conductivity by a factor of around 10^4. Even so, the actual values of conductivity tend to remain quite small. For NaCl, for example, the temperature must be around 500 °C (773 K) in order to achieve a conductivity (about $10^{-5}\,\Omega^{-1}\,m^{-1}$) comparable with that of deionised water at room temperature!

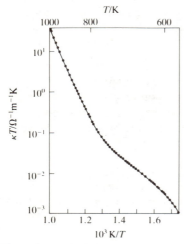

Figure 36 The ionic conductivity of NaCl plotted against reciprocal temperature.

The relatively recent surge of interest in ionic conductors stems from the discovery of materials where the conductivity exceeds $1\,\Omega^{-1}\,m^{-1}$ at temperatures below their melting points, due to the passage through the solid of just one particular type of ion. In Section 5.6.1 and 5.6.2, you will see that such materials have commercial applications as **solid electrolytes** in electrochemical cells. To fulfill such a role, an ionic solid must not only have a high ionic conductivity but also have negligible *electronic* conductivity. A cell is a device for converting chemical energy *directly* to electricity. It does this by ensuring that the complementary oxidation and reduction processes (equations 4 and 5 in Section 4.2, for example) that together make up the cell reaction occur at *separate* sites—the two electrodes. Thereby, the electrons 'exchanged' in the reaction are forced to travel through the external circuit. If the electrolyte shows a substantial electronic conductivity, it can short-circuit this process, leading to a wasteful self-discharge of the cell.

At present, only a small fraction of ionic solids come up to these rather searching standards. These are often referred to as **fast ion conductors**. To understand some of the features of structure and chemical composition that help to promote such fast conduction, we shall start by looking more closely at the way ions move through a crystal lattice.

5.2 Ionic conductivity in solids

We might expect the conductivity of ionic solids to be poor. The pictures we use to represent such solids show the situation in a pure, perfect crystal: every regular lattice site is occupied by the correct type of ion. It's difficult to envisage how an ion could make its way through such a highly ordered array. In fact, ionic conductivity in solids like NaCl is only possible because real crystals *always* contain *defects*. The important ones for our present purposes are *point defects*—individual sites in the lattice where an ion is either missing completely, or else displaced into an interstitial position.

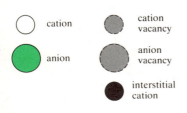

Figure 37 Schematic representation of the vacancy mechanism.

We shall look at the different types of defect more closely in a moment. But the important point is that their presence creates a situation much more favourable for ionic motion. For example, the presence of vacancies immediately suggests the possibility of a rather simple mechanism, the **vacancy mechanism**, whereby an ion 'hops' or 'jumps' from a regular lattice site to an adjacent vacant one: this is illustrated schematically in Figure 37. Notice that the mechanism can be seen in two equivalent ways: either the ion jumps to the right, or the vacancy moves to the left. This is important, because the formal definition of ionic conductivity, κ, is:

$$\kappa = cZeu \qquad\qquad\qquad \textbf{8}$$

Here c is the concentration of charge carriers (usually expressed as the number per unit volume); Ze is their charge (expressed as the appropriate multiple of the electronic charge, $e = 1.602\,189 \times 10^{-19}\,\mathrm{C}$); and u is their **mobility**, which is a measure of the drift velocity (Section 2.5) in a constant electric field.

This definition holds for *all* types of electrolyte, but the above discussion suggests that we must be careful when applying it to solid electrolytes. In particular, it is the defects that effectively determine how many ions in a solid are 'free to move', so c is identified with the *concentration of defects*. Equally, u is the mobility of the ions by virtue of these defects.

Although the mechanism sketched in Figure 37 is grossly simplified (the counterions in the lattice are not even shown!), it is sufficient to suggest that ionic conductivity in solids can be enhanced in two ways: by increasing the concentration of defects, or the speed at which they migrate through the crystal—or both. We shall look at each factor in turn.

5.3 Defects and their concentration

Point defects fall into two broad categories: intrinsic defects, which are integral to the crystal in question and do not change its overall composition—whence the alternative name, **stoichiometric defects**; and extrinsic defects, which are created by inserting foreign ions into the lattice.

5.3.1 Intrinsic defects

SLC 3 As you may recall from Second Level, intrinsic defects are themselves of two types: these are shown for a slice through a crystal of composition MX in Figure 38. In this case, a *Schottky defect* (Figure 38a) consists of a *pair* of vacant sites: a *cation vacancy* which is compensated by an *anion vacancy* in order to preserve charge neutrality.

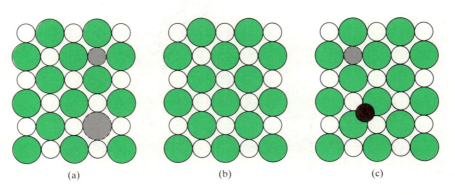

○ cation   cation vacancy

● anion   anion vacancy

● interstitial cation

Figure 38 Schematic illustration of intrinsic point defects in a crystal of composition MX: (a) Schottky pair; (b) perfect crystal; (c) Frenkel pair.

(a)   (b)   (c)

☐   How would you describe a Schottky defect in an MX_2 type structure?

■   Now each cation (M^{2+}) vacancy will require two anion (X^-) vacancies to balance the charges—a Schottky trio.

In fact, simple Schottky defects are more common in 1 : 1 compounds: examples include NaCl itself, ZnS (wurtzite) and CsCl.

Frenkel defects also involve vacancies in the structure, but this time they usually occur on only one **sublattice** of the crystal. A *cation Frenkel defect* is shown schematically in Figure 38c: it involves a cation moving into an interstitial position (*not* a regular lattice site), thereby creating a cation vacancy. Such defects are particularly important in crystals of the light-sensitive silver halides (especially AgBr) used in photography. Here, their presence is believed to play a vital part in the formation of the photographic image.

☐   It is less common to find *anion Frenkel defects*, where anions are displaced into interstitial sites. Can you suggest a reason for this?

■   In general, the anions in an ionic structure are *larger* than the cations (only four simple cations—Rb^+, Cs^+, Tl^+ and Ba^{2+}—are larger than the smallest anion, F^-), so it is usually more difficult for them to enter (often rather crowded) interstitial sites.

Important exceptions to this general rule are to be found among compounds that crystallise with the fluorite structure. This group includes a number of fluorides (MF_2—CaF_2 itself, for example, SrF_2 and PbF_2) and oxides (MO_2—ThO_2, UO_2, ZrO_2) that are reasonably good anionic conductors—under some conditions at least. A notable example is PbF_2, where the conductivity in the solid state reaches a value (around $10^2 \, \Omega^{-1} \, m^{-1}$ at 720 K (*ca.* 450 °C)) comparable with that found in the melt.

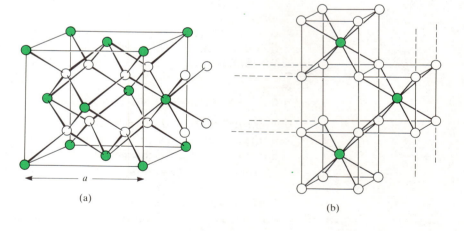

(a)

(b)

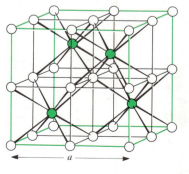

(c)

(d)

Figure 39 The crystal structure of fluorite MX_2. (a) Unit cell as an f.c.c. array of cations; (b) and (c), the same structure redrawn as a simple cubic array of anions: the unit cell is marked by a green outline in (c); (d) cell dimensions.

● cation   ○ anion

So, what is so special about the fluorite structure? The ideal structure (Figure 39a) is based on a cubic close-packed array of cations, with all the tetrahedral holes occupied by anions: the crucial point is that this leaves *all* the larger octahedral holes vacant. As usually presented, this arrangement produces a face-centred unit cell, with an empty octahedral site at the body centre (Figure 39a). A different, and perhaps clearer, view of it can be made by noting that the anions form a simple cubic array—the unit cell is now formed by eight small cubes or **octants**, with the centre of each alternate octant occupied by a cation (Figure 39b): the unit cell can then be represented as shown in Figure 39c. The two views are completely equivalent: either description highlights the openness of the fluorite structure, and points to a possible candidate for the interstitial position where a Frenkel defect forms—the normally vacant octahedral site, which is also the centre of an empty cube of anions.

5.3.2 The energy of defect formation

Assuming this to be the case suggests a rather simple way of obtaining a crude estimate of the energy for defect formation in a fluorite structure. Let's assume this to be a halide, MX_2. We start by estimating, and comparing, the potential energy of an anion (X^-) on a regular anion site with one on the proposed interstitial position. To do this, we restrict attention to the coulombic interaction with nearest neighbours and next-nearest neighbours—and use the expression for

SLC 9

the potential energy of two ions that was introduced at Second Level:

$$\text{potential energy} = -\frac{e^2 Z}{4\pi\varepsilon_0 r}$$

$$= -(2.31 \times 10^{-28}\,\text{J m})Z/r \qquad \textbf{9}$$

where we have introduced the values of the physical constants e and ε_0. Here, Z is the charge on the other ion (so $Z = -1$ for interaction with another anion in the structure, or $+2$ for interaction with a cation) and r is the distance between them.

If the face-centred cubic unit cell dimension in Figure 39a is a, the cube side in Figure 39b is $0.5a$: simple geometry then shows that the distance from a regular anion site to the cube centre is $\sqrt{3}a/4 = 0.43a$ (Figure 39d). With this background, you should be able to make the estimate for yourself: *do it now by working through the following SAQs.*

SAQ 17 (*revision*) Use Figure 39 to describe the coordination by nearest neighbours and next-nearest neighbours of an anion: (a) on a regular lattice site; and (b) on the proposed interstitial position at the cube centre.

SAQ 18 Restricting attention to the interactions identified in SAQ 17, use equation **9** to obtain an estimate of the energy of defect formation in fluorite itself, where $a = 537\,\text{pm}$.

Some experimental *enthalpies of defect formation* are collected in Table 5 (*overleaf*); clearly your estimate is not at all bad! Given the simplicity of our model*, this measure of agreement is undoubtedly somewhat fortuitous. But working through the calculation does bring to light an important general point: the energy of an ion in a given structure depends—often in a very sensitive way—on its precise location within the lattice. Not only does this determine the energy of defect formation, but it also maps out the energy landscape that an ion must cross in order to move through the crystal. As we shall see shortly, it is this landscape that controls the mobility of the ion.

5.3.3 The concentration of defects

The values in Table 5 also suggest a general way of manipulating the population of intrinsic defects, whether they are Schottky or Frenkel, or even a mixture of the two. Notice the sign of ΔH: defect formation is always an *endothermic* process.

---

* Among other things, our crude model makes no allowance for more distant interactions, or for internuclear repulsion (other than in treating ions as hard spheres), or lattice vibrations—or any changes to the lattice in response to defect formation (lattice relaxation, so called)! All of these, and other factors can be taken into account in sophisticated computer models.

Table 5 Enthalpies of formation of Schottky and Frenkel defects in selected compounds

	Compound	$\Delta H/10^{-19}$ J	ΔH/eV
Schottky defects	MgO	10.57	6.6
	CaO	9.77	6.1
	LiF	3.749	2.34
	LiCl	3.397	2.12
	LiBr	2.88	1.8
	LiI	2.08	1.3
	NaCl	3.685	2.30
	KCl	3.621	2.26
Frenkel defects	UO_2	5.45	3.4
	ZrO_2	6.57	4.1
	CaF_2	4.49	2.8
	SrF_2	1.12	0.7
	AgCl	2.56	1.6
	AgBr	1.92	1.2
	β-AgI	1.12	0.7

☐ If you think of defect formation as a pseudo-chemical reaction (analogous perhaps to the endothermic dissolution of a sparingly soluble electrolyte), what effect will a rise in temperature have on the concentration of defects?

SFC 4 ■ It should increase. According to Le Chatelier's principle, for an endothermic process, increasing the temperature favours the forward reaction—defect formation in this case—so the number of defects should increase.

This analogy with chemical equilibrium can be elaborated on to derive an expression for the number of intrinsic defects in a crystal of a given type (MX or MX_2 or M_2X, or whatever). Here, we simply quote the results of that analysis: both expressions relate to a solid of simple MX-type stoichiometry.

Thus, the *concentration of Schottky defects* (c_S)—the number per unit volume—is given by:*

$$c_S = n \exp(-\Delta H_S/2kT) \qquad \qquad \textbf{10}$$

where n is the number of 'normal' cation or anion sites per unit volume of the crystal and k is the Boltzmann constant ($k = 1.380\,662 \times 10^{-23}$ J K^{-1}). Similarly, for Frenkel defects, the analogous expression is:

$$c_F = (nn_i)^{1/2} \exp(-\Delta H_F/2kT) \qquad \qquad \textbf{11}$$

where n and n_i are now the concentrations of lattice sites and potential interstitial sites, respectively.

SAQ 19 To get a feel for the numbers involved, and the effect of temperature, take a typical value of ΔH, say 5×10^{-19} J (Table 5). Then use equation 10 to calculate the value of c_S/n—the proportion of empty sites—at two temperatures: room temperature, say 300 K (27 °C), and 1 000 K. Now repeat the calculations with $\Delta H = 10^{-19}$ J.

For a typical ionic solid, the concentration of intrinsic defects can obviously be expected to be very low at ambient temperatures. Raising the temperature certainly has a dramatic effect on the defect population. Even at 1 000 K, however, there are still fewer than one per million lattice sites.

But there is another way of increasing the numbers of defects: reduce the energy of defect formation. In practice, it is difficult to see quite how the value of ΔH in

* If you are unfamiliar with the *exponential function* exp (x) or e^x, then you should note that e is a number (e ≈ 2.718 282) that is the base of *natural logarithms* (ln)—just as 10 is the base of *common logarithms* (log). Thus, if you take the natural logarithm of an exponential e^x (exp (x)), the answer is x, that is ln e^x = x; or equally ln e^{-x} = $-x$. To get some practice at working out exponentials, make sure you try SAQ 19.

a given solid could be manipulated in this way. But perhaps we can find solids in which ΔH is lower than average—that is, solids where there are available sites in the structure with energies comparable to those of the regular lattice sites: in this case, the defect concentration and hence the ionic conductivity should be high. If you think back to the calculation in SAQ 18, this suggests that the ideal situation is one in which there is a built-in excess of sites *identical* to the normally occupied positions, so that $\Delta H = 0$ (since e^{-x} approaches 1 as x becomes smaller and smaller—try a few values on your calculator to convince yourself of this). As you might expect, such structures are rare, because ions are generally packed closely together in order to maximise the coulombic interactions. But examples are known—and we shall look at one of them, α-AgI, in Section 5.5.

SAQ 20 Unlike fluorides, *pure* oxides with the fluorite structure show high anion conduction only at elevated temperature—above 2 300 K or so. Can you suggest a possible reason for this?

5.3.4 Extrinsic defects

Another—and more generally applicable—strategy for increasing the defect population is to create them, by **doping** the parent crystal with selected impurities. A technologically important example is **zirconia, ZrO_2**. When pure, this has the fluorite structure only at high temperatures, but the structure can be stabilised at room temperature by adding the oxides of divalent or trivalent metals—commonly CaO or Y_2O_3.

☐   Given that structural studies show that the added Ca^{2+} ions replace Zr^{4+} ions on the face-centred cubic lattice, what effect would you expect the addition of CaO to have on the defect population of ZrO_2?

■   Simple substitution of cations would upset the overall charge neutrality of the crystal. In this case, it can be restored by creating vacancies on the oxide sublattice—one for each Ca^{2+} ion incorporated (Figure 40).

In principle, other mechanisms of achieving charge compensation are possible: we shall look at some of them, and their consequences, later in the Block. In practice, however, *doping with cations of lower charge* usually does produce anion vacancies—and the effect can be dramatic. For example, **stabilised zirconias** can be made with proportions of CaO at the 10–20 mole per cent level! The vast defect population that this implies is sufficient to induce considerable oxide ion mobility at technologically useful temperatures—around 1 000 K, or so.

SAQ 21 In NaCl, the cations are more mobile than the anions. Assuming a simple substitution mechanism, what effect (if any) would you expect small amounts of the following impurities to have on the conductivity of NaCl: (a) AgCl; (b) $CaCl_2$; (c) NaBr; (d) Na_2O?

5.4 Ionic mobility

Whatever their origin, the point defects that influence the ionic conductivity of a solid are of only two types—vacancies or interstitial ions. This suggests a second simple mechanism to add to the **vacancy mechanism** sketched earlier (Figure 37, Section 5.2); it is shown in Figure 41, and termed (not surprisingly!) the **interstitial mechanism**. Other, more complex, mechanisms can be envisaged—involving cooperation between vacancies and interstitials, for example—but these simple pictures are sufficient to illustrate the model of ionic motion we shall adopt. It is called the **hopping model**. As the name suggests, it describes ionic motion as a series of jumps from site to site—whether it is lattice site to vacancy, or interstitial position to interstitial position, or whatever.

5.4.1 The model

To develop the model, it helps to look at a three-dimensional representation of the jump. Suppose the crystal has a sodium chloride structure, with vacancies on the cationic sublattice, as shown in Figure 42a. Imagine now that the central

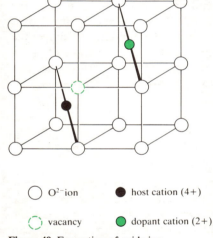

○ O^{2-} ion   ●   host cation (4+)

◌   vacancy   ●   dopant cation (2+)

Figure 40 Formation of oxide ion vacancies by doping ZrO_2 with CaO.

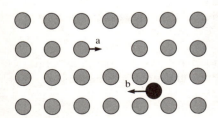

Figure 41 Schematic representation of ionic motion by (a) a vacancy mechanism, or (b) an interstitial mechanism.

cation (shown black) jumps into the adjacent vacant site. To see what this entails, we extract the coordination octahedron of the filled site, rotate this a little for clarity, and then draw in the coordination octahedron of the empty site alongside it. (The end result (Figure 42b) reminds us that the NaCl structure can be thought of as octahedra joined edge to edge—and connected, via shared triangular faces (shaded green), by tetrahedra.)

Figure 42b also reveals that the vacancy mechanism may be a good deal less simple than it appears in a two-dimensional drawing. Notice in particular that the jump can trace out a number of routes: two of them are marked by broken lines in the Figure.

SAQ 22 Describe the coordination by anions at each of the points labelled 1–4 in Figure 42b.

The crucial point is that either of these routes—or any other path—will take the the ion into environments where its energy is necessarily *higher* than it is on a regular lattice site. This is the **energy landscape** referred to in Section 5.3.2. In general, the ion will follow the lowest energy path available to it, and a typical, but highly schematic, *energy profile* is shown in Figure 43.

Pictures like this draw on analogies with the energy profiles for elementary *chemical reactions* introduced in the Science Foundation Course. The most obvious difference is that the beginning and end of the reaction are now at the same energy—for a jump like the one in Figure 42, at least, where the ion moves from one site to an equivalent one. But here, as in the Foundation Course, the energy labelled E_a is called the *activation energy* and can be thought of as an **energy barrier** to the process in question: in this case, the ion must acquire this energy in order to overcome the strong interactions that it encounters *en route*.

5.4.2 Developing the model
As before, this analogy with chemical reactions can be elaborated into a quantitative model—capable of making predictions that can be tested against experiment. We can only outline the steps involved here.

The starting point recognises that ions are not at rest on their lattice sites: they are vibrating. A vibration in the direction of the jump—toward the face of the octahedron in Figure 42b, say—can then be seen as an attempt to make the jump. The success rate will depend both on the **attempt frequency**, v—which can be identified with the frequency of the lattice vibration (typically 10^{12}–10^{13} Hz)—and the height of the energy barrier, E_a. Arguments like this lead to the following expression for the **jump frequency**—the rate at which successful jumps are made.

$$\text{jump frequency} = v \exp\left(-E_a/kT\right) \qquad \textbf{12}$$

The exponential term in this expression tells us the probability that the jump will be successful.

☐   Does the form of this term seem reasonable?

■   Yes. As you saw in Section 5.3.3, the smaller x is, the closer to unity e^{-x} becomes: in this case, the lower the energy barrier, E_a, is, the more likely the jump is to succeed.

The mobility, u, of the ions under the influence of an electric field is proportional to the jump frequency. This gives us an expression of the following simple form:

$$u = A \exp\left(-E_a/kT\right) \qquad \textbf{13}$$

Here A is a constant (in the sense that it does not depend on temperature)*, but it incorporates quantities like the attempt frequency and other parameters depen-

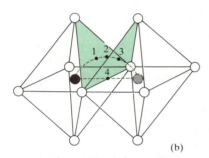

○ X^-

● M^+

◕ M^+ vacancy

SFC 5

Figure 42 Sodium chloride type structure, showing (a) coordination octahedron of central cation, (b) coordination octahedra of central cation and adjacent vacancy.

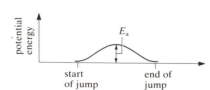

Figure 43 Schematic representation of the change in energy during motion of an ion along the lowest energy path.

* This is not strictly true, but the temperature dependence of this term is usually much weaker than that in the exponential term.

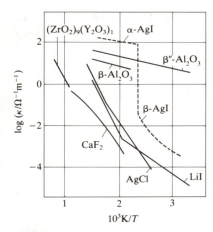

Figure 44 The conductivities of selected solid electrolytes over a range of temperatures.

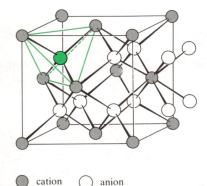

cation ⊙ anion

Figure 45 The fluorite structure, showing the coordination tetrahedron (green) around one of the anions (also green).

(a)

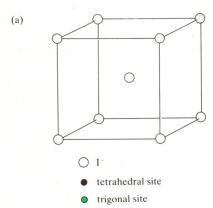

○ I^-
● tetrahedral site
● trigonal site

(b)

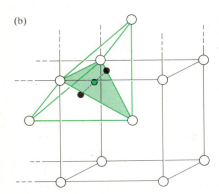

Figure 46 Building up the structure of α-AgI: (a) The body-centred cubic array of I^- ions. (b) The positions of two tetrahedral sites and a trigonal site between them (green).

dent on the ion and lattice in question. To complete the analysis, we combine this result with our expression for ionic conductivity (equation 8, Section 5.2) to give:

$$\kappa = cZeA \exp\left(-E_a/kT\right)$$

or $\kappa = A'c \exp\left(-E_a/kT\right)$ **14**

where we have swept the charge (Ze) into the constant term.

This simple expression accounts for the increase of ionic conductivity with temperature, and predicts that plots like those in Figure 44 will be linear. As you can see they generally are (ignoring the very odd plot for α-AgI for the moment). In fact, equation 14 can also account for the curiously bent shapes of plots for certain solids (like LiI, for example), but we shall not go into that here. Suffice it to say that the comparison with experiment supports the essential features of our model. It also provides a route to experimental values of the activation energy. Typical values of E_a for ionic solids lie in the range 0.05–1.1 eV—somewhat lower than the energies of defect formation.

Notice the parallels between equation 13 and our expressions for the concentration of defects—be they Schottky (equation 10) or Frenkel (equation 11). Once again, the energy requirement and the temperature turn up in an exponential term, so we would expect our earlier conclusions to apply. Thus, ionic mobility can obviously be enhanced by raising the temperature. But the thrust of most developments in this field is to achieve high conductivity at moderate temperatures—and this requires a low energy barrier. Indeed, the activation energy is the parameter that best characterises a solid electrolyte—a good electrolyte usually being reckoned to have E_a less than about 0.2 eV.

So, can we identify factors that help to minimise the energy barrier—and hence add to the list of requirements for fast conduction that we started in Section 5.3.4? As suggested there, we shall look for insights to one of the best ionic conductors known, α-AgI.

SAQ 23   Refer again to Figure 42b. Which of the two pathways marked there would you expect to present the higher energy barrier?

SAQ 24   Figure 45 is the fluorite structure again—but drawn to show the coordination tetrahedron of the anion coloured green. (The anion behind this tetrahedron has been omitted for clarity.) Suppose the coloured anion jumps to the octahedral hole at the body centre. Sketch, and describe, the journey it takes in terms of the changing coordination by cations.

5.5   α-AgI: a study in fast ion conduction

At room temperature, silver iodide has a conductivity of around $2 \times 10^{-4}\,\Omega^{-1}\,m^{-1}$, a reasonably high—if not remarkable—value for an ionic solid. As shown in Figure 44, the conductivity increases fairly smoothly with rising temperature, until 419 K (146 °C)—where it suddenly shoots up by a factor of some 10^4! This dramatic rise marks a *phase transition*. Below 419 K, there are two phases of AgI: γ-AgI which has the zinc blende structure, and β-AgI with the wurtzite structure. Both of these are based on a *close-packed array* of iodide ions—cubic in one case (γ-AgI) and hexagonal in the other (β-AgI).

By contrast, the high-temperature phase—known as α-AgI—has a structure based around a *body-centred cubic* array of iodide ions, as shown in Figure 46a.

☐   How many iodide ions are there in the unit cell?

■   Two. ($8 \times \frac{1}{8}$) at the corners, plus 1 at the body-centre.

So the unit cell has to accommodate two Ag^+ ions— but where, exactly? Careful inspection of the structure in Figure 46a reveals an extraordinary number, and variety, of possibilities. The first step is to recognise that a body-centred cubic structure—*unlike a face-centred cubic* (or indeed hexagonal-cubic packed) *structure*—can be thought of as built up exclusively from distorted tetrahedra,

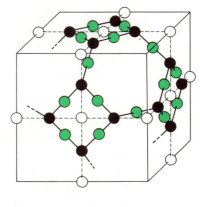

● tetrahedral site

● trigonal site

○ distorted octahedral site

Figure 47 Possible cation sites in the b.c.c. structure of α-AgI. The thick solid and broken lines mark possible diffusion paths.

joined by sharing common triangular faces. Two adjacent 'tetrahedra' are marked out in Figure 46b. The four-coordinate positions at the centre of each 'tetrahedron' fall on the faces of the body-centred cubic unit cell—four to each face forming the pattern of sites shown in Figure 47: despite the slight distortion, we shall refer to these as tetrahedral sites. (Here we have omitted the I^- ions for clarity. Ignore the thick lines and broken lines for the moment.)

☐ How many tetrahedral sites are there in each unit cell?

■ 12. $(4 \times 6 \times \frac{1}{2})$ since each face is shared between two unit cells.

As you found in answering SAQs 22 and 24, the centre of each triangular face is a three-coordinate site—or **trigonal site**. In this structure, these fall between the tetrahedral sites, both on the cell faces and *between* pairs of adjacent faces—as indicated in Figure 47: there are 24 such sites per unit cell. Also marked on this figure are sites that have a six-coordinate, octahedral geometry—albeit quite highly distorted. There are six of these per unit cell, making a total of 42 possible sites for our two Ag^+ ions. Some choice!

☐ Would you expect all of these sites to be equally attractive to the Ag^+ ions?

■ No. The whole thrust of our discussion has been that the energy of an ion depends in a sensitive way on its immediate environment—particularly, the coordination geometry of that environment.

In this case, careful structural studies on single crystals of α-AgI indicate that the octahedral sites are the smallest and most crowded. They appear to be empty, with the Ag^+ ions largely (but not exclusively, as we shall see) distributed among the 12 tetrahedral sites—*all of which have the same energy*. This means that an excess of five available, *and equivalent*, sites per Ag^+ ion are built in to the structure—confirmation that α-AgI amply fulfills the criterion regarding the defect concentration that we posited in Section 5.3.3.

But what of the **conduction path**—the route that gives α-AgI an experimental activation energy of just 0.05 eV? This is believed to be *via a trigonal site*—as indicated by the solid lines in Figure 47: effectively the Ag^+ ion jumps *directly* from one 'tetrahedral' lattice site into the neighbouring one by passing through the fairly open tetrahedral face. Calculations confirm that this results in a rather flat energy profile—with an energy barrier of just 0.05 eV, in complete agreement with the experimental value. By contrast, jumps via the constricted octahedral sites (along the fine broken lines in Figure 47) must surmount a barrier of some 0.12 eV.

In fact, these same studies effectively map out the diffusion paths, showing the electron density of the Ag^+ ions to be somewhat spread out along them. The interpretation is that the structure is actually a very dynamic one, with any given Ag^+ ion hardly pausing between jumps—continually moving from one of the tetrahedral sites to another. This more liquid-like behaviour has sometimes been described as a **molten sublattice** of Ag^+ ions.

A flat energy profile means that the energy of the moving ion does not change much during its journey, so what general lessons does α-AgI offer about ways to achieve this?

Firstly, notice that the coordination changes relatively little (from 4 to 3, and back to 4). As here, most crystal structures will usually force the ion to pass through a site of low coordination at some point. For example, you saw in answering SAQs 22 and 24 that movement through any face-centred cubic structure will almost certainly start with a jump through a trigonal position. It follows that we might expect *ionic mobility to be higher for ions of low coordination number in their regular (or interstitial) sites*. In fact, several silver(I) and copper(I) compounds that show exceptionally high cationic conductivity all have either body-centred cubic or close-packed frameworks, with the cations in tetrahedral sites; some examples are given in Table 6.

Table 6 'Disordered' cation compounds related to α-AgI

Body-centred cubic	Face-centred cubic	Hexagonal close-packed
α-AgI	α-Ag$_2$HgI$_4$	β-CuCl
α-Ag$_3$SI	α-CuI	β-Cu$_2$S
α-Ag$_2$S	α-Cu$_2$HgI$_4$	
α-Ag$_2$Se		
α-CuBr		

A second, rather obvious, factor is the *charge* on the ion: the smaller this is, the smaller will any changes be in the coulombic interactions experienced by the ion as it moves. Thus, *monovalent ions are more likely to have high mobility*. Most of the known fast ion conductors do, in fact, involve monovalent ions.

Finally, a less obvious point: Ag$^+$ and I$^-$ are both said to be rather *polarisable ions*—they fall into the soft category (of acids and bases, respectively) that you met earlier in the Course. Such ions behave somewhat less like hard spheres than most: we can think of them as a bit malleable—able to adjust their electron distribution to differing environments. The net effect is to lessen changes in the interaction energy along the migration path, so *polarisable ions tend to enhance the ionic mobility*.

It is unusual to find all these features in a given compound as they are in α-AgI—and none of them can be divorced from the determining effect of the underlying crystal structure. Nevertheless, they do provide a framework within which to discuss the conductivities of compounds with related structures and compositions. *The following group of SAQs provides some examples for you to work through.*

SAQ 25 The structure of γ-AgI is shown in Figure 48, and has a face-centred cubic array of one type of ion, I$^-$, with Ag$^+$ in *half* of the tetrahedral holes. Describe briefly the major similarities and differences between this structure and that of α-AgI. To what do you ascribe the lower conductivity of Ag$^+$ ions in the low-temperature form?

SAQ 26 Most of the examples in Table 6 contain either iodide or sulphide (or a heavier member of Group VI) as the anionic sublattice. How does the presence of these anions contribute to the high cationic conductivity of these compounds?

SAQ 27 Can you now suggest further reasons why pure oxides with the fluorite structure are poorer anionic conductors than fluorides? (Refer back to SAQ 20, Section 5.3.3, if necessary.)

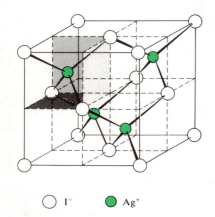

○ I$^-$ ● Ag$^+$

Figure 48 The zinc blende structure of γ-AgI.

5.6 Solid electrolytes in action

The solids we discussed in the last Section all show high conductivity by virtue of their intrinsic structure and composition. We close this part of the Block by looking at two applications involving solid electrolytes that might be termed extrinsic—in the sense that their conductivity can be attributed to defects induced by impurities. In the first case—**stabilised zirconias**—doping induces vast numbers of *vacancies* spread throughout the oxide ion sublattice—giving enhanced oxide ion conductivity.

In our second example—a crystalline oxide known as **β-alumina***—the structure accommodates large numbers of interstitial Na$^+$ ions in the lattice, and their almost unimpeded motion in certain layers gives the solid a sodium-ion conductivity close to liquid values.

5.6.1 Oxygen meters
A battery harnesses the energy of a spontaneous chemical reaction—and translates it into a voltage between the terminals of the cell. In much the same way, a **concentration meter** (often called a **sensor**) can harness the tendency for a substance to equalise its concentration—or pressure, for a gas—across a permeable

---

* . . . somewhat incongruously, it transpires! The compound was originally thought to be a polymorph of alumina, Al$_2$O$_3$—and was named as such. It was only later found to contain Na$^+$ ions.

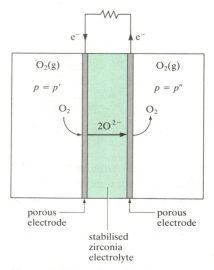

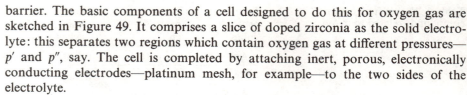

Figure 49 Schematic representation of an oxygen meter.

barrier. The basic components of a cell designed to do this for oxygen gas are sketched in Figure 49. It comprises a slice of doped zirconia as the solid electrolyte: this separates two regions which contain oxygen gas at different pressures—p' and p'', say. The cell is completed by attaching inert, porous, electronically conducting electrodes—platinum mesh, for example—to the two sides of the electrolyte.

Suppose $p' > p''$, so there is a tendency for oxygen to move from the region on the left to that on the right of our Figure. Connecting up the cell as shown allows this to happen *electrochemically*. Thus, oxygen gas is reduced to oxide ions at the left-hand electrode (LHE). The ions migrate through the electrolyte, to be reoxidised to oxygen gas at the right-hand electrode (RHE):

LHE:	$O_2(\text{g at } p') + 4e \longrightarrow 2O^{2-}$	**15**
RHE:	$2O^{2-} \longrightarrow O_2(\text{g at } p'') + 4e$	**16**
overall:	$O_2(\text{g at } p') \longrightarrow O_2(\text{g at } p'')$	**17**

Earlier in the Course, you met the *Nernst equation*, and saw how it expresses the composition dependence of individual electrode potentials as:

$$E = E^{\ominus} - \left\{\frac{2.303RT}{nF}\right\} \log \left\{\frac{a_X{}^x a_Y{}^y \cdots}{a_A{}^a a_B{}^b \cdots}\right\} \qquad \textbf{18}$$

This expression can be applied equally well to an **overall cell reaction**—like the one in equation 17: the value of E is then a measure of the voltage delivered by the cell, and n is the number of electrons transferred in the electrode reactions (equations 15 and 16).

SAQ 28 (*revision*) Now apply equation 18 to the reaction in equation 17.

To use the cell in Figure 49 as an **oxygen meter**, one of the pressures (p'', say) is held constant—usually by using a reference gas of known pressure (p_{ref}). Depending on the application, this reference may be pure oxygen—or just the atmosphere (with $p_{O_2} \approx 21\,\text{kPa}$ (0.21 atm)). With this modification, the expression you derived becomes:

$$E = \left(\frac{2.303RT}{4F}\right) \log (p'/p_{ref}) \qquad \textbf{19}$$

Evidently, the voltage recorded is a direct measure of the desired pressure—*provided the electrolyte shows negligible electronic conductivity* (as mentioned in Section 5.1). In fact, you will see shortly that *non-stoichiometric* oxides are often markedly electron conducting—and stabilised zirconias are no exception. Fortunately, in this case the electronic contribution depends on the oxygen pressure, and becomes significant only at pressures *below* about $10^{-11}\,\text{Pa}$. Thus, the use of zirconia-based cells to monitor and control oxygen pressures above this value is already well established.

The system is particularly useful for remote sensing of oxygen gas in hostile environments such as furnaces and flues of many kinds. For example, it can be used as an **oxygen probe** in the exhaust gas stream from a vehicle engine (Figure 50). Feedback from the probe allows the air/fuel mixture to be adjusted to the optimum value for efficient combustion—thereby both saving energy and reducing air pollution. Probes can also be designed to measure the oxygen content of molten metals (steel, for example)—often a crucial parameter in determining the quality of the end-product.

The zirconia used in these meters is stabilised with fairly modest levels of dopant (typically around the 10 mole per cent level), the most commonly used oxides being CaO and Y_2O_3. Higher dopant levels produce such massive concentrations of defects that strong interactions can arise between them—leading to new 'ordered' arrangements within the crystal. You will meet examples of such structural features later in the Block. The crucial point here is that any order of this kind reduces the conductivity. This is important because even optimum doping

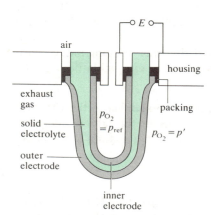

Figure 50 Sketch of a zirconia probe that could be used to measure the pressure of oxygen in exhaust gases. (The housing and packing must be electronic conductors.)

levels induce only a fairly modest conductivity (in the $1{-}10\,\Omega^{-1}\,m^{-1}$ range at $1\,300\,K$ (*ca.* $1\,000\,°C$), such that the use of oxygen meters is presently restricted to temperatures above $1\,000\,K$ (*ca.* $700\,°C$) or so.

5.6.2 β-alumina and the sodium–sulphur battery

Potentially important is the use of β-alumina as the electrolyte in a high-energy, lightweight, rechargeable battery that is likely to prove a powerful competitor to the lithium–titanium disulphide system described in Section 4.2. The battery is based on the **sodium–sulphur battery**, first described by N. Weber and J. T. Kummer of the Ford Motor Company in 1967. The driving force behind this development was almost certainly the discovery of the remarkably high conductivity of Na^+ ions (some $10{-}100\,\Omega^{-1}\,m^{-1}$ at $620\,K$ (*ca.* $350\,°C$) in β-alumina. And to understand that, we need to look at the underlying crystal structure.

In fact, the name β-alumina covers a family of related materials. Here we start with the parent compound, a sodium aluminium oxide represented by the ideal formula $Na_2O\,.\,11Al_2O_3$. The key to its structure is a repeating pattern of slabs containing cubic close-packed layers of oxide ions, alternating with single low-density layers where three-quarters of the oxygen is missing.

As shown in Figure 51a, the slabs are actually just four layers thick. Within these slabs, the Al^{3+} ions occupy both octahedral and tetrahedral holes between the oxygen layers, in such a way that the structure resembles a slice of the mineral *spinel* ($MgAl_2O_4$) which you met earlier in the Course—but without the Mg^{2+} ions. So, a fully occupied oxygen layer *plus* the aluminium below it has the formula Al_3O_4. The quartets of close-packed oxygen layers are separated by a quarter-full oxygen layer, and to bind the oxygens of this defective layer to the quartet on each side requires an aluminium above and an aluminium below (Figure 51b). To give a charge balance, every oxygen in the defective layer is matched by a sodium. From the ratios of atoms in Figure 51b, we can see that this gives an overall stoichiometry of $NaAl_{11}O_{17}$. The Na^+ ions are found, *and move*, exclusively within the open, oxygen-deficient layers—the **conduction planes**.

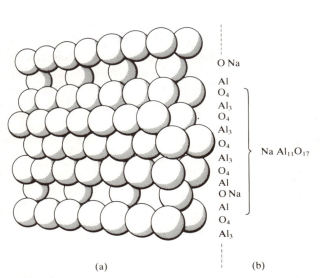

(a)

(b)

Figure 51 (a) Oxide layers in β-alumina. (b) The ratio of atoms in each layer of the structure.

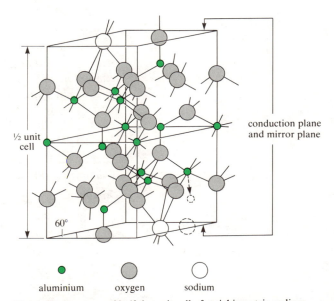

aluminium   oxygen   sodium

Figure 52 Structure of half the unit cell of stoichiometric sodium β-alumina. (The broken lines indicate the effect of incorporating additional oxygen in the conduction plane.)

X-ray studies provide the structure shown in Figure 52 (ignore the broken lines for the moment). Notice that the conduction plane is also a mirror plane, so the second half of the unit cell is just a reflection of the first.

Close inspection of the conduction plane (Figure 53) reveals a situation not unlike the one we met earlier with α-AgI. In particular, the Na^+ ions (one for each oxide ion in the plane, Figure 52) have a variety of sites open to them,

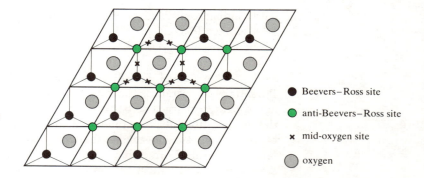

Figure 53 Schematic view of the geometry of the conduction plane in sodium β-alumina. The mid-oxygen sites lie midway between each pair of Beevers–Ross and anti-Beevers–Ross sites. The fine lines indicate the Na^+ diffusion network.

● Beevers–Ross site

● anti-Beevers–Ross site

× mid-oxygen site

○ oxygen

known as the **Beevers–Ross (BR) site**, **anti-Beevers–Ross (aBR) site**, and **mid-oxygen (mO) site**; as shown in Figure 53, there are five sites for each Na^+ ion.

Calculations suggest that the BR sites are the most stable, and this is reflected in the structure of pure *stoichiometric* β-alumina. Unlike α-AgI, it seems to be highly ordered, with almost exclusive occupancy of the BR sites. In fact, this material is rather unstable, but indications are that it has a fairly poor ionic conductivity—with an activation energy in excess of 0.6 eV.

As normally prepared, β-aluminas are *always* non-stoichiometric, being considerably *richer* in sodium than the ideal formula. This has a dramatic effect on the conductivity. For example, materials of practical interest usually have an increase in sodium content of 20–30 mole per cent, with activation energies as low as 0.15 eV. Here, every two additional Na^+ ions are compensated by an extra oxide ion—placed at, or near, the mO sites in the conduction planes (Figure 53). They are locked at these positions by Al^{3+} ions displaced from their regular sites within the spinel blocks—as indicated by the broken lines in Figure 52.

The additional Na^+ ions join the pool already present in the conduction planes; structural studies indicate that the pool content is now mobile over the various sites, along the honeycomb marked by fine lines in Figure 53—a further similarity with α-AgI.

Crystals yet richer in sodium usually fall into a second, but closely related, family of structures—the β″-aluminas—which differ in the stacking of the close-packed slabs. In general, these materials have higher conductivities (see Table 7, for example), but they tend to be less stable and more sensitive to moisture. In practice, a compromise is adopted, and crystals with both types of structure—β and β″—are usually present in the strong, fine-grained ceramic employed as electrolyte in sodium–sulphur cells; typical conductivities are in the range $10–20\,\Omega^{-1}\,m^{-1}$ at 620 K (*ca.* 350 °C).

Table 7 Ionic conductivities and activation energies of typical β-alumina and β″-alumina

Specimen		κ (at 573 K)/$\Omega^{-1}\,m^{-1}$	E_a/eV
β	single crystal	21.3	0.13
	polycrystalline	8.0	0.25
β″	single crystal	100	0.10
	polycrystalline	20–40	0.16–0.22

In most cell designs, the electrolyte is formed into a tube, which also serves to contain one of the reactants—*liquid sulphur* in the example sketched in Figure 54. Because this has a high resistance, it must be dispersed in an electronically conducting matrix—generally carbon felt or mat—and provided with a *current collector* to make contact with the external circuit. The other reactant is *molten sodium*; the circuit is completed via the outer cell case.

During *discharge*, sodium is oxidised to Na^+ ions, which travel through the electrolyte—to the inner interface in our sample cell. Here, they take part in

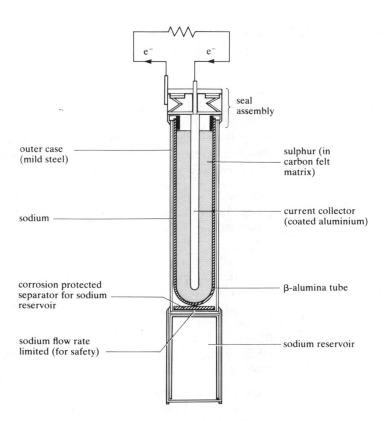

Figure 54 Schematic representation of a 'central sulphur' cell. Typically, the electrolyte tube is some 300 mm long and 30 mm in diameter.

rather complex reactions, the net effect being the reduction of sulphur to sodium *polysulphides*. In the early stages of discharge, the reactions can be represented as follows:

$$2Na(l) = 2Na^+ + 2e \qquad\qquad \textbf{20}$$

$$\underline{2Na^+ + 5S(l) + 2e = Na_2S_5(l)} \qquad\qquad \textbf{21}$$

$$\overline{2Na(l) + 5S(l) = Na_2S_5(l)} \qquad\qquad \textbf{22}$$

Later on, lower polysulphides are formed, and the discharge is usually terminated when the melt reaches a composition equivalent to Na_2S_3—thereby ensuring that the mixture remains molten at the normal operating temperature (around 620 K). Despite their seeming complexity, the electrode processes can be reversed successfully, and cells capable of hundreds of discharge/recharge cycles have already been built.

As before (see Section 4.2), interest in the sodium–sulphur system stems from its theoretical **energy density**: put simply, it involves a high-energy reaction (equation 22) between relatively light substances. In practice, energy densities have been achieved some 5 or 6 times better than those delivered by commercially available lead–acid batteries. No wonder the system is such an attractive candidate for powering electric vehicles—and it is undergoing extensive development and testing in this context.

One final point: at first sight, the idea of a battery containing quantities of molten sodium and sulphur at high temperatures is alarming—even more so, in view of the extremely corrosive nature of the polysulphide melt. In fact, awareness of the potential hazards has led to safety features being incorporated at every stage in the development of the sodium–sulphur system, and it seems likely that its commercial exploitation is just around the corner.

SAQ 29 *Undoped* β-alumina shows a maximum conductivity and minimum activation energy when the sodium excess is around 20–30 mole per cent. Thereafter, further increase in the sodium content causes the conductivity to decrease. By contrast, β-alumina crystals *doped* with Mg^{2+} have a much higher conductivity than do undoped crystals. Can you explain these observations?

6 NON-STOICHIOMETRIC COMPOUNDS

We saw in the last Section that it is possible to *introduce* defects into a perfect crystal by adding an impurity. Such an addition causes point defects of one sort or another to form, but they no longer occur in complementary pairs: such impurity-induced defects are said to be *extrinsic*. When assessing what defects have been created in a crystal, an important point to remember is that *the overall charge on the crystal must always be zero*.

SAQ 30 If a crystal of NaCl is heated in sodium vapour, sodium is taken into the crystal, and the formula becomes $Na_{1+x}Cl$. What are the possible structures of this new compound?

The solid referred to in SAQ 30 is called a **non-stoichiometric compound**, because the ratios of the atomic components are no longer the simple integers that we have come to expect for well characterised compounds. A careful analysis of many substances—particularly inorganic solids—shows that it is not uncommon for the atomic ratios to be irrational: uranium dioxide for instance can range in composition from $UO_{1.65}$ to $UO_{2.25}$—certainly not the perfect UO_2 that we might expect! There are many other examples, some of which we will go on to explore.

So, what kind of compounds are likely to be non-stoichiometric? 'Normal' covalent compounds have a fixed composition, and the atoms are usually held together by strong covalent bonds formed by the pairing of two electrons. Breaking these bonds usually takes quite a lot of energy, and under normal circumstances, a particular compound does not show a range of composition; this is true for most organic compounds, for instance. Ionic compounds also are usually stoichiometric, because to remove or add ions requires a great deal of energy. We have seen, however, that it is possible to make ionic crystals non-stoichiometric by doping them with an impurity, as with the example of Na added to NaCl in SAQ 30. There is also another mechanism whereby ionic crystals can become non-stoichiometric: if the crystal contains an element that can form ions of more than one charge, then a change in the number of those ions can be compensated by changes in ion charge; this maintains the charge balance but alters the stoichiometry.

☐   Where would we look for ions that behave in such a way?

■   As in much of this Course, to the transition elements and to the lanthanides and actinides.

In summary, non-stoichiometric compounds have formulae that do not have simple integer ratios of atoms; they also usually exhibit a range of composition (as exemplified by UO_2). They can be made by introducing impurities into a system, but are frequently a consequence of the ability of the metal component to exhibit variable valency. Table 8 lists a few non-stoichiometric compounds together with their composition ranges.

Until recently, defects both in stoichiometric and non-stoichiometric crystals have been treated entirely from the point of view that point defects were randomly distributed. However, research is now showing that isolated point defects are *not* scattered at random in non-stoichiometric compounds but are often dispersed throughout the structure in some kind of regular pattern. This Section largely tries to explore the relationship between stoichiometry and structure.

The structure determination of compounds containing defects is a very difficult problem, and it is only very recently that much of our knowledge has been gleaned. Why is it so difficult? Diffraction methods as they are normally used yield an *average* structure for a crystal. For pure, relatively defect-free structures this is a good representation, but for non-stoichiometric and defect structures it avoids precisely the information that you want to know. For this kind of struc-

Table 8 Approximate composition ranges for some non-stoichiometric compounds

Compound		Composition range*
TiO_x	$[\approx TiO]$	$0.65 < x < 1.25$
	$[\approx TiO_2]$	$1.998 < x < 2.000$
VO_x	$[\approx VO]$	$0.79 < x < 1.29$
Mn_xO	$[\approx MnO]$	$0.848 < x < 1.00$
Fe_xO	$[\approx FeO]$	$0.833 < x < 0.957$
Co_xO	$[\approx CoO]$	$0.988 < x < 1.000$
Ni_xO	$[\approx NiO]$	$0.999 < x < 1.000$
CeO_x	$[\approx Ce_2O_3]$	$1.50 < x < 1.52$
ZrO_x	$[\approx ZrO_2]$	$1.700 < x < 2.004$
UO_x	$[\approx UO_2]$	$1.65 < x < 2.25$
$Li_xV_2O_5$		$0.2 < x < 0.33$
Li_xWO_3		$0 < x < 0.50$
TiS_x	$[\approx TiS]$	$0.971 < x < 1.064$
Nb_xS	$[\approx NbS]$	$0.92 < x < 1.00$
Y_xSe	$[\approx YSe]$	$1.00 < x < 1.33$
V_xTe_2	$[\approx VTe_2]$	$1.03 < x < 1.14$

* Note that all composition ranges are temperature dependent and the figures here are intended only as a guide.

ture determination, a technique that is sensitive to *local* structure is needed, and such techniques are very scarce. Structures are often elucidated from a variety of sources of evidence: X-ray and neutron diffraction, density measurements, spectroscopy (when applicable) and more recently high-resolution electron microscopy (HREM); magnetic measurements have also proved useful in the case of FeO. Probably HREM has done most to clarify the understanding of these structures, as it is capable under favourable circumstances of giving information on an atomic scale by direct lattice imaging.

We should not leave this Section without noting that non-stoichiometric compounds are of potential use to industry because their electronic, optical, magnetic and mechanical properties can be modified by changing the proportions of the atomic constituents. This is widely exploited and researched by the electronics industry.

6.1   Non-stoichiometry in wustite, FeO

Ferrous oxide is known as **wustite**, **FeO** and it has the NaCl crystal structure. Accurate chemical analysis shows that it is non-stoichiometric: it is always deficient in iron.

☐   There are two ways in which an iron deficiency could be accommodated by a defect structure. Can you think what they are?

■   There could be *iron vacancies*, giving a formula $Fe_{1-x}O$; or alternatively there could be an *excess of oxygen in interstitial positions*, with the formula FeO_{1+x}.

A comparison of the theoretical and measured *densities* of the crystal can distinguish between the alternatives; let's see how to go about such a calculation. First of all, there is a certain amount of experimental data to be collected and assembled. We need to know the experimentally measured density of the crystal: the easiest method of measuring the density of a crystal is the **flotation method**. Liquids of differing densities (that dissolve in one another) are mixed together until a mixture is found that will *just* suspend the crystal—it neither floats nor sinks. The density of that liquid mixture must then be the same as that of the crystal, and it can be found by weighing an accurately measured volume.

Now for the *theoretical* density of the crystal. This can be obtained from the volume of the unit cell and the mass of the unit cell contents. The results of an

X-ray diffraction structure determination would give us both of these data, as the unit cell dimensions are accurately measured and the type and number of atoms in the unit cell are also found. We'll work through an example of this type of calculation now for FeO.

A particular crystal of FeO was found to have a unit cell dimension of 430.1 pm, a measured density of $5.728 \times 10^3\,kg\,m^{-3}$, and an iron to oxygen ratio of 0.945.

☐ What is the volume of the unit cell?

■ FeO has the NaCl structure, which is cubic—all sides are equal in length and all angles are 90°—the volume is thus $(430.1\,pm)^3 = 7.956 \times 10^7$ $(pm)^3 = 7.956 \times 10^{-29}\,m^3$.

☐ How many formula units of FeO are there in a unit cell?

■ Four.

How can we calculate the mass of four FeO formula units? The relative atomic mass of Fe is 55.85 and of O is 16.00, and so one mole of FeO weighs $(55.85 + 16.00)\,g = 71.85\,g$ or $0.071\,85\,kg$.

☐ How do we get from here to the mass of one formula unit?

■ We must divide by Avogadro's number ($N_A = 6.022\,045 \times 10^{23}\,mol^{-1}$).

Now calculate the mass of four FeO formula units, which is the mass of a unit cell containing no defects. The mass is given by $(4 \times 0.071\,85)/N_A\,kg = 4.773 \times 10^{-25}\,kg$.

Because of the non-stoichiometry, the sample we are considering doesn't have *exactly* four formula units in the unit cell, so let's first take the case of *iron vacancies*. We know from the chemical analysis data that the Fe : O ratio is 0.945, so on average, each unit cell will contain four oxygens and (4×0.945) or 3.780 irons.

☐ What will be the mass of the contents of this unit cell?

■ $[(3.780 \times 55.85) + (4 \times 16.00)]/(N_A \times 10^3)\,kg$.

☐ Use this expression now to calculate the density.

■ You should have divided the mass by the volume $7.956 \times 10^{-29}\,m^3$, and found the density to be $5.742 \times 10^3\,kg\,m^{-3}$.

Turning our attention now to the *interstitial* case:

☐ What is the ratio of O to Fe atoms in the unit cell?

■ O : Fe will be given by $1/0.945 = 1.058$.

This unit cell will, on average, contain four irons and (4×1.058) or 4.232 oxygens. Its mass will be

$$[(4 \times 55.85) + (4.232 \times 16.00)]/(N_A \times 10^3)\,kg.$$

Using this to calculate density, we get $6.076 \times 10^3\,kg\,m^{-3}$. Comparing the results, it is clear that the experimentally measured density ($5.728 \times 10^3\,kg\,m^{-3}$) is very close in value to that calculated for a cell containing *iron vacancies* ($5.742 \times 10^3\,kg\,m^{-3}$), and we should therefore write the formula as $Fe_{0.945}O$.

SAQ 31 Confirm the results given above for a different sample of wustite which has a unit cell dimension of 428.2 pm, an Fe : O ratio of 0.910 and an experimental density of $5.613 \times 10^3\,kg\,m^{-3}$.

Table 9 Experimental and theoretical densities (10^3 kg m^{-3}) for FeO

O:Fe ratio	Fe:O ratio	Lattice parameter*/pm	Observed density	Theoretical density	
				interstitial O	Fe vacancies
1.058	0.945	430.1	5.728	6.076	5.742
1.075	0.930	429.2	5.658	6.136	5.706
1.087	0.920	428.5	5.624	6.181	5.687
1.099	0.910	428.2	5.613	6.210	5.652

* Unit cell dimension

Using the results from the above examples and adding further data, we can draw up a table of densities for FeO (Table 9).

It is found to be characteristic of all non-stoichiometric compounds that *their unit cell size varies smoothly with composition but the symmetry is unchanged*. This is known as **Vegard's law**.

We have now found two characteristic properties of non-stoichiometric compounds; first, that they exist over a range of composition, and second, that throughout that range the unit cell length varies smoothly with no change of symmetry. We have also found that it is possible to determine whether the non-stoichiometry is accommodated by vacancy or interstitial defects using density measurements.

SAQ 32 How does the change in lattice parameter of FeO with iron content corroborate the iron-vacancy model and refute an oxide-interstitial model?

6.1.1 Electronic defects in FeO
The discussion of the defects in FeO has so far been only structural. Now we turn our attention to the balancing of the charges within the crystal. In principle, the compensation for the iron deficiency can be made either by oxidation of some iron(II) ions *or* by reduction of some oxide anions. It is energetically more favourable to oxidise iron(II).

☐   How will electrical neutrality be maintained?

■   For each Fe^{2+} vacancy, *two* Fe^{2+} cations must be oxidised to Fe^{3+}.

In the overwhelming majority of cases, defect creation involves changes in the cation oxidation state.

☐   How would you expect electrical neutrality to be maintained in the case of metal excess in simple oxides?

■   Neighbouring cation(s) would normally be *reduced*.

In a later Section, we will look at some general cases of non-stoichiometry in simple oxides, but before we do that we will complete the FeO story with a look at its detailed structure.

6.1.2 The structure of FeO
Let's recap what we already know. FeO has the NaCl structure with Fe^{2+} ions in octahedral sites. The iron deficiency manifests itself as cation vacancies, and the electronic compensation made for this is that for every Fe^{2+} ion vacancy there are two neighbouring Fe^{3+} ions.

We might reasonably expect that the Fe^{2+}, Fe^{3+}, and the cation vacancies are randomly distributed over the octahedral sites in the cubic close-packed O^{2-} array in wustite. Structural studies (X-ray studies, and neutron diffraction and magnetic measurements) have shown that this is *not* the case and that *some of the Fe^{3+} ions are in tetrahedral sites*.

Although the structure of wustite is still under debate, it appears that it contains various types of **defect cluster**, and that these new structures are distributed

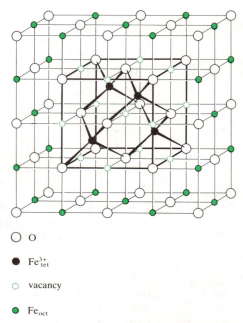

O O

● Fe^{3+}_{tet}

○ vacancy

● Fe_{oct}

Figure 55 The Koch–Cohen cluster shown with the front and back planes cut away for clarity. The central section with the four tetrahedrally coordinated Fe^{3+} ions is picked out in bold.

throughout the crystal. A defect cluster is a region of the crystal where the defects form an *ordered structure*. One possibility is known as the **Koch–Cohen cluster**, and part of it is shown in Figure 55. It bears a strong resemblance to the structure of Fe_3O_4, the next highest oxide of iron. We can think of wustite, therefore, as fragments of Fe_3O_4 intergrown in the NaCl structure of FeO. (The crystal structure of Fe_3O_4 is discussed in a different context later in the Block in Section 7 on magnetism.)

At the centre of the Koch–Cohen cluster is a modified FeO unit cell—drawn in bold in the Figure. This contains *four additional tetrahedrally coordinated* Fe^{3+} *ions* in the tetrahedral holes; furthermore, the octahedrally coordinated iron sites at the centre of this unit cell, and at the mid-points of its edges, are vacant. The other octahedrally coordinated iron sites in the cluster are occupied, but as you will see in a moment, they may contain either iron(II) or iron(III) ions. We therefore designate them simply as Fe_{oct}. *The front and back planes have been cut away from the diagram in Figure 55, to make the central section more visible.*

☐ How many NaCl-type unit cells make up the whole cluster?

■ Allowing for the missing front and back layers there are 8—2 × 2 × 2.

In the central (bold) unit cell, there are vacancies at *all* the octahedral sites, which would normally be occupied by Fe^{2+} ions.

☐ Defect clusters in FeO are often referred to by the ratio of cation vacancies to interstitial Fe^{3+} in tetrahedral sites. What is the ratio for the Koch–Cohen cluster?

■ 13 : 4; there are 12 vacancies at the mid-points of the edges of the central, bold unit cell, and one at its centre, making 13 in all.

Figure 55 does not show a complete face of a Koch–Cohen cluster. Try to draw one for yourself now and answer the following questions before looking at Figure 56.

☐ What type of atoms lie at the corners of the face? How many Fe_{oct} ions are there totally enclosed within the boundaries of each face? What type of atom lies at the mid-point of the cell edges?

- Fe_{oct} ions lie at the corners of the faces and at the mid-points of the edges. There are, in addition, *five* Fe_{oct} ions contained within the boundary of each face.

☐ How many oxide ions does a cluster contain?

■ The cluster is composed of eight NaCl-type unit cells. There are no anion defects, so there are $(8 \times 4) = 32$ oxide ions.

We are now in a position to determine the contents of the cluster, using the same methods as for a normal unit cell. There are eight Fe_{oct} ions at the corners, contributing $(8 \times \frac{1}{8}) = 1\ Fe_{oct}$.

☐ What are the contributions to Fe_{oct} from the cell edges and faces?

■ The 12 mid-point Fe_{oct} ions each contribute $\frac{1}{4}$, giving 3 Fe_{oct}. There are $(6 \times 5) = 30\ Fe_{oct}$ ions bounded within the faces, each of which contribute $\frac{1}{2}$ to the cluster, making 15 Fe_{oct}.

These, together with the four tetrahedral Fe^{3+}, which are completely contained within the cluster, make a grand total of 23 Fe cations. The overall formula for the cluster is thus, $Fe_{23}O_{32}$—almost Fe_3O_4!

Having determined the atomic contents of the cluster, we must now turn our attention to the charges. There are 32 oxide ions, so the Fe cations overall must have *64* positive charges.

☐ How many are accounted for by the Fe^{3+} ions in tetrahedral positions?

■ 12, leaving 52 to find from the remaining 19 Fe_{oct} cations.

This accounting is difficult to do by inspection, so we need to employ some simple mathematics. Suppose that there are $x\ Fe^{2+}$ ions and $y\ Fe^{3+}$ ions in octahedral sites, we know that

$$x + y = 19 \qquad\qquad 23$$

We also know that their total charges must equal 52, so:

$$2x + 3y = 52 \qquad\qquad 24$$

giving two simultaneous equations. Substituting $x = 19 - y$ in equation 24, gives $y = 14$ and thus $x = 5$.

The octahedral sites surrounding the central (bold) unit cell are thus occupied by 5 Fe^{2+} ions and by 14 Fe^{3+} ions. By injecting such clusters throughout the FeO structure, the non-stoichiometric structure is built up. The exact formula of the compound (the value of x in $Fe_{1-x}O$) will depend on the average separation of the injected clusters.

SAQ 33 Figure 57 shows just the central section of another possible defect cluster for FeO. Determine (a) the vacancy : interstitial ratio for this cluster; (b) assuming that this section is surrounded by Fe ions and oxide ions in octahedral sites as in the Koch–Cohen cluster, determine the formula of a sample made totally of such clusters; and (c) determine the numbers of Fe^{2+} and Fe^{3+} ions in octahedral sites.

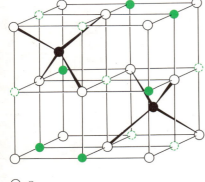

○ O

◌ vacancy

● Fe_{oct}

● Fe^{3+}_{tet}

Figure 57 A possible cluster in $Fe_{1-x}O$.

6.2 Uranium dioxide

In Sections 6.1–6.1.2, you saw how a metal-deficient oxide was achieved through cation vacancies. We now turn to an example where metal deficiency arises from interstitial anions.

Above $1\,400\,K$ $(1\,127\,°C)$, a single oxygen-rich non-stoichiometric phase of UO_2 is found with formula UO_{2+x}, ranging from UO_2 to $UO_{2.25}$.

☐ What formula does $UO_{2.25}$ correspond to?

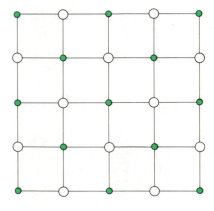

○ O

● Fe$_{oct}$

Figure 56 One face of the Koch–Cohen cluster.

■ U$_4$O$_9$, which is a well characterised oxide of uranium.

UO$_2$ has the *fluorite structure*. This is illustrated in Figure 58 (ignore the green lines for the time being).

☐ How many formula units of UO$_2$ does the unit cell in Figure 58a contain?

■ 4. There are 4 uranium ions contained within the cell boundaries. The 8 oxide ions come from: $(8 \times \frac{1}{8}) = 1$ at the corners, $(6 \times \frac{1}{2}) = 3$ at the face centres, $(12 \times \frac{1}{4}) = 3$ at the cell edges, and 1 at the cell mid-point.

As more oxygen is taken into UO$_2$, the extra oxide ions go into interstitial positions.

☐ From Figure 58a, where is the most obvious interstitial position?

■ In the centre of one of the octants with no metal ion.

In fact, neutron diffraction shows that an interstitial oxide anion does not sit exactly in the centre of an octant but is displaced sideways; this has the effect of moving *two* other oxide ions from their lattice positions by a very small amount, leaving two vacant lattice positions. This is illustrated in Figure 58, where in (a) three vacant octants are picked out in green, and in (b) the positions of the one additional interstitial oxide and the two displaced oxides with their vacancies are shown. The movement of the ions from 'ideal' positions is shown by small arrows: the movement of the interstitial oxide from the centre of an octant is along the direction of a diagonal of one of the cube faces, whereas the displacement of the oxide ions on lattice positions is along cube diagonals.

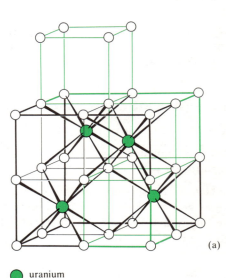
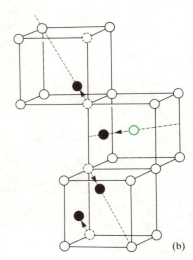

(a)   (b)

● uranium

○ oxygen

○ ideal interstitial site for oxygen

● interstitial oxygen

○ vacancy

Figure 58 (a) The fluorite structure of UO$_2$ with a unit cell marked in bold and the defect cluster position in green. (b) Interstitial defect cluster in UO$_{2+x}$. Uranium positions (not shown) are in the centre of every other cube.

☐ What is the atomic composition of the unit cell in Figure 58a when modified by the defect structure shown in Figure 58b?

■ U$_4$O$_9$; the loss of the displaced lattice oxygen at the centre of the top face is balanced by the gain of that at the centre of the bottom face. The net oxygen gain is just the single new interstitial in the central octant of Figure 58b.

U_4O_9 is found to be the oxygen-rich limit for the UO_2 non-stoichiometric structure. We can think of UO_{2+x}, therefore, as containing **microdomains** of the U_4O_9 structure within that of UO_2.

☐ How do you expect compensation to be made for the negative charges on the extra interstitial oxide anions?

■ Most likely, neighbouring uranium(IV) atoms will be oxidised to either uranium(V) or uranium(VI).

The story as we have told it above is actually a slightly simplified version of the truth. There are *two* different (but similar) positions that the interstitial oxides can take within the structure, and when these are ordered throughout the structure, we find a very large unit cell for U_4O_9 based on $4 \times 4 \times 4$ fluorite unit cells (with a volume of 64 times that of the UO_2 unit cell). However, in this Course we shall not go into the details of this!

6.3 The TiO structure

Titanium and oxygen form non-stoichiometric phases, which exist over a range of composition centred about the stoichiometric $1:1$ value, from $TiO_{0.65}$ to $TiO_{1.25}$. We shall just look at what happens in the upper range, from $TiO_{1.00}$ to $TiO_{1.25}$.

As you saw earlier in the Block, at the composition $TiO_{1.00}$, the crystal structure is based on a defective NaCl structure: one-sixth of the titaniums, and one-sixth of the oxygens are missing. In fact, above $1\,170\,K$ (*ca.* 900 °C) these vacancies are *randomly* distributed, but, as you saw in Section 4, below this temperature they are *ordered* as shown in Figure 28. Figure 28 also showed that this ordering gives rise to a monoclinic unit cell. Let us use it to confirm the fact that TiO is a one-sixth defective NaCl structure.

☐ Consider first the titanium vacancies in Figure 28, and ask yourself the following questions. How many titanium vacancies are there at the corners, the edges, the faces and contained within the cell?

■ There are vacancies only at the corners, $(8 \times \frac{1}{8}) = 1$, and two of the faces $(2 \times \frac{1}{2}) = 1$.

☐ Now ask yourself the same questions for the titanium ions.

■ Cell edges, $(4 \times \frac{1}{4}) = 1$. Cell faces, $(8 \times \frac{1}{2}) = 4$ on the top and bottom. There are 5 ions contained within cell boundary, making 10 in total.

The titanium stoichiometry of the unit cell that we have worked out is obviously representative of the whole structure: of the 12 sites, 10 are occupied and 2 are vacant. You can also use Figure 28 to convince yourself that this is true for oxygen. As we discussed in Section 4, it is this vacancy structure that allows sufficient contraction of the lattice, and therefore overlap of d orbitals, to allow metallic conductivity.

When titanium oxide has the formula $TiO_{1.25}$, it has a different defect structure, still based on the NaCl structure, but with all the oxygens present, and one in every five titaniums missing. The pattern of the titanium vacancies is shown in Figure 59, which shows a layer of the type in Figure 28 with the oxygens omitted; only titaniums are marked. If you draw horizontal lines through the titaniums, you will see that every fifth one is missing. Where samples of titanium oxide have formulae that lie between the two limits discussed here, $TiO_{1.00}$ and $TiO_{1.25}$, the structure seems to consist of portions of the $TiO_{1.00}$ and $TiO_{1.25}$ structures intergrown.

SAQ 34 Figure 60 shows layers in the $TiO_{1.25}$ structure with both the Ti and O sites. Use this and Figure 59 to demonstrate that the unit cell shown has the correct stoichiometry for the crystal.

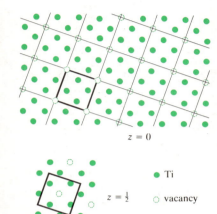

Figure 59 Successive Ti layers in the structure of $TiO_{1.25}$.

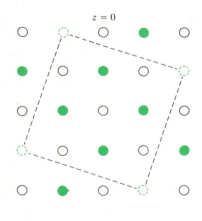

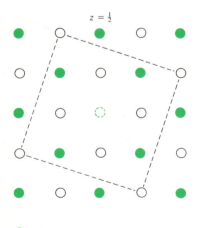

Figure 60 The structure of $TiO_{1.25}$ showing both O and Ti positions.

6.4 Crystallographic shear

Non-stoichiometric compounds are found for the higher oxides of tungsten, molybdenum and titanium, WO_{3-x}, MoO_{3-x} and TiO_{2-x}, respectively. The reaction of these systems to the presence of point defects is entirely different from what we've met previously; in fact, the point defects are *eliminated* by a process known as **crystallographic shear (CS)**.

In these systems, a series of closely related compounds with very similar formulae and structure exists. The formulae of these compounds all obey a *general formula*, which for the molybdenum and tungsten oxides is Mo_nO_{3n-1} and W_nO_{3n-1} respectively, and for titanium dioxide is Ti_nO_{2n-1}; n can vary, taking values of 4 and above. The resulting series of oxides is known as a **homologous series**.

☐ List the first eight members of the molybdenum trioxide series.

■ Mo_4O_{11}, Mo_5O_{14}, Mo_6O_{17}, Mo_7O_{20}, Mo_8O_{23}, Mo_9O_{26}, $Mo_{10}O_{29}$ and $Mo_{11}O_{32}$.

In these compounds, we find regions of corner-linked octahedra of one structure separated from each other by thin regions of a different structure: these regions are known as the **crystallographic shear planes**. The different members of a homologous series are determined by the fixed spacing between the CS planes. The structure of a shear plane is quite difficult to understand and these structures are usually depicted by the linking of octahedra as described in Section 1.

SAQ 35 Take a very simple case where *two* metal oxide octahedra eliminate oxygen by sharing. How does the formula change as they (a) share a corner, (b) share an edge, and (c) share a face?

Now let us turn to one of the non-stoichiometric compounds that we are interested in, WO_{3-x}. The WO_3 structure is the same as that of ReO_3, which is shown in Figure 61 (ignore the bold squares for the time being). ReO_3 is made up of $[ReO_6]$ octahedra that are linked together via their corners—each corner of an octahedron is shared with another. Figure 61a shows part of one layer of the octahedra in the structure. Notice that, *within the layer*, any octahedron is linked to four others; it is also linked, via its upper and lower corners, to octahedra in the layers above and below.

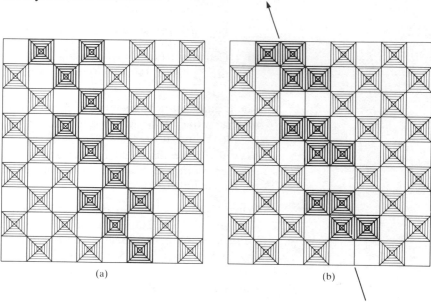

Figure 61 Formation of shear structure.

(a) (b)

☐ Justify how the linking of $[ReO_6]$ octahedra gives the overall formula ReO_3. (It helps to draw a line diagram or use your model kit to see the linkages.)

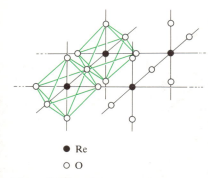

Figure 62 Part of the ReO_3 structure showing the linking of octahedra through the corners.

● Re
○ O

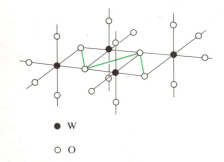

● W
○ O

Figure 63 Group of four $[WO_6]$ octahedra sharing edges (marked in green) formed by the creation of shear planes in W_nO_{3n-1}.

■ Part of the ReO_3 structure is drawn in Figure 62, and you can see that every oxygen atom is shared between two metal atoms. As six oxygens surround each Re, the overall formula is ReO_3.

The non-stoichiometry in WO_{3-x} is achieved by some of the octahedra in the structure changing from corner sharing to edge sharing. Look back now to the octahedra marked in bold in Figure 61a. The edge sharing corresponds to shearing the structure so that the chains of bold octahedra are displaced to the positions in Figure 61b. This shearing occurs at regular intervals in the structure and is interspersed with slabs of the 'ReO$_3$' structure (corner-linked $[WO_6]$ octahedra). It creates concentrates of four octahedra that share edges. The direction of maximum density of the edge-sharing groups is called the crystallographic shear plane and is indicated by an arrow in Figure 61b.

Now, let's see how this alters the stoichiometry of WO_3. To do this, we need to find the stoichiometry of one of the groups of four octahedra that are linked together by sharing edges.

□ Try drawing a line diagram of this group of four octahedra (or making a model) and finding its stoichiometry.

■ A drawing of one of these groups is shown in Figure 63.

The group in Figure 63 consists of 4 W atoms and 18 O atoms. 14 of the O atoms are linked out to other octahedra (these bonds are indicated) so are each shared by 2 W atoms, while the remaining 4 O atoms are only involved in the edge sharing within the group. The overall stoichiometry will be given by $[4W + (14 \times \frac{1}{2})O + 4O]$, giving W_4O_{11}.

Clearly, if groups of four octahedra with stoichiometry W_4O_{11} are interspersed throughout a perfect WO_3 structure, then the amount of oxygen in the structure is reduced and we can write the formula as WO_{3-x}. Now let's be a little more precise about this and see what is the quantitive effect of introducing the groups of four in an ordered way.

First of all, suppose that the structure sheared in such a way that the entire structure was composed of these groups—the formula would then become W_4O_{11}. Now, let there be one $[WO_6]$ octahedron for each group of four—the overall formula now becomes $[W_4O_{11} + WO_3] = W_5O_{14}$.

□ What will be the formulae of structures that contain 2, 3, 4, 5, 6 and 7 $[WO_6]$ octahedra per group of four?

■ $W_4O_{11} + 2WO_3 = W_6O_{17}$; $W_4O_{11} + 3WO_3 = W_7O_{20}$;
$W_4O_{11} + 4WO_3 = W_8O_{23}$; $W_4O_{11} + 5WO_3 = W_9O_{26}$;
$W_4O_{11} + 6WO_3 = W_{10}O_{29}$; $W_4O_{11} + 7WO_3 = W_{11}O_{32}$.

Can you see a pattern emerging here? Take the basic formula of the group of four, W_4O_{11}, and set $n = 4$.

□ What is 11 in terms of n?

■ $3n - 1$, when $n = 4$.

Does this hold for all the other formulae that you have worked out? Try it and see. The answer is yes it does. So you can see that we have produced the formula for the homologous series simply by introducing set ratios of the edge-sharing groups in among the $[WO_6]$ octahedra.

The shear planes are found to repeat throughout a particular structure in a regular and ordered fashion, so any particular sample of WO_{3-x} will have a specific formula corresponding to one of those listed above. The different

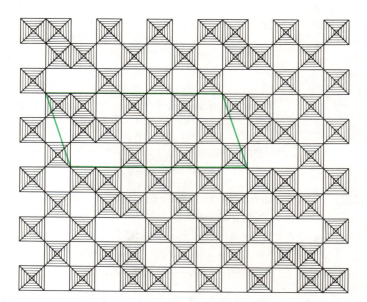

Figure 64 A member of the W_nO_{3n-1} homologous series with a unit cell marked.

members of the homologous series are determined by the fixed spacing between the CS planes. An example of one of the structures is shown in Figure 64. A unit cell has been marked to help you count the ratio of $[WO_6]$ octahedra to the groups of four.

☐ Can you see what the formula of the structure is in Figure 64?

■ Within the marked unit cell, there is one group of four and seven octahedra giving the overall formula $W_4O_{11} + 7WO_3 = W_{11}O_{32}$.

Members of the Mo_nO_{3n-1} series have the same structure as their W_nO_{3n-1} analogue, even though unreduced MoO_3 doesn't have the ReO_3 structure, but a layer structure.

SAQ 36 Figure 65 shows another member of the homologous series, W_nO_{3n-1}. What formula does it correspond to?

Figure 65 A member of the W_nO_{3n-1} homologous series.

6.5 Electronic properties of non-stoichiometric oxides

Earlier, we considered the structure of non-stoichiometric FeO in some detail. If we apply the same principles to other binary oxides, we can define four types of compound:

Metal excess
Type A: anion vacancies present; formula MO_{1-x}
Type B: interstitial cations; formula $M_{1+x}O$

Metal deficiency
Type C: interstitial anions; formula MO_{1+x}
Type D: cation vacancies; formula $M_{1-x}O$

Figure 66 Structural possibilities for binary oxides.

(a) Type A oxides: Metal excess/anion vacancies. (i) This shows the two electrons that maintain charge neutrality, localised at the vacancy. (ii) The electrons are associated with the normal cations making them into M^+.

(b) Type B oxides: Metal excess/interstitials. (i) This shows an interstitial *atom*, whereas in (ii) the atom has ionised to M^{2+}, and the two liberated electrons are now associated with two normal cations, reducing them to M^+.

(c) Type C oxides: Metal deficiency/interstitial anions. The charge compensation for an interstitial anion is by way of two M^{3+} ions.

(d) Type D oxides: Metal deficiency/cation vacancies. The cation vacancy is compensated by two M^{3+} cations.

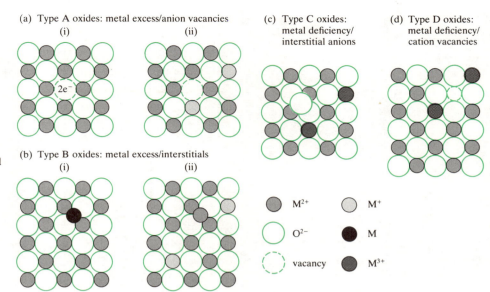

(a) Type A oxides: metal excess/anion vacancies

(i) (ii)

(b) Type B oxides: metal excess/interstitials

(i) (ii)

(c) Type C oxides: metal deficiency/interstitial anions

(d) Type D oxides: metal deficiency/cation vacancies

M^{2+} M^+ O^{2-} M   vacancy M^{3+}

Figure 66 illustrates some of the structural possibilities for simple oxides with type A, B, C and D non-stoichiometry, assuming that they have the NaCl structure.

☐ Into which category does FeO fall?

■ FeO is **type D**. (Other compounds falling into the type D category are MnO, CoO and NiO.)

Type A oxides compensate for metal excess with **anion vacancies**. To maintain the overall neutrality of the crystal, two electrons have to be introduced for each anion vacancy. These can be trapped at a vacant anion site, as is shown in (i). However, it is an extremely energetic process to introduce electrons into the crystal, and so we are more likely to find them associated with the metal cations as is shown in (ii), which we can describe as reducing those cations from M^{2+} to M^+.

Type B oxides have a metal excess which is incorporated into the lattice in **interstitial** positions. This is shown in (i) as an interstitial atom, but it is more likely that the situation in (ii) will hold, where the interstitial atom has ionised and the two electrons so released are now associated with two neighbouring ions, reducing them from M^{2+} to M^+. Cadmium oxide, CdO, has this type of structure. Zinc(II) oxide is also type B, but has the wurtzite structure.

Type C oxides compensate for the lack of metal with **interstitial anions**. The charge balance is maintained by the creation of two M^{3+} ions for each interstitial anion, each of which we can think of as M^{2+} associated with a positive hole.

Before considering the conductivity of these non-stoichiometric oxides, it is probably helpful to recap on what we know about the structure and properties of the stoichiometric binary oxides of the first-row transition elements. A summary of the properties of binary oxides is given in Table 10 (*overleaf*).

We discussed the conductivity of the stoichiometric oxides in Section 4.3, and saw that their conductivity is dependent on two competing effects: on the one hand, the d orbitals overlap to give a band—the bigger the overlap, the greater the band width—and electrons in the band are delocalised over the whole structure; on the other hand, interelectronic repulsion tends to keep electrons localised on individual atoms. TiO and VO behave as *metallic conductors* and must therefore have good overlap of the d orbitals producing a d electron band. We saw that this overlap arises partly because Ti and V are early in the transition series (the d orbitals have not suffered the contraction, due to increased nuclear charge, seen later in the series), and partly because of the unusual crystal structure, where $\frac{1}{6}$ of the titaniums and oxygens (VO is similar) are missing from an NaCl-type structure, allowing contraction of the structure and thus better d orbital overlap.

54

Table 10 Properties of the first-row transition element monoxides

Element	Ca	Sc	Ti	V	Cr	Mn	Fe	Co	Ni	Cu	Zn
Structure of stoichiometric oxide MO	NaCl structure	doesn't exist	defect NaCl $\frac{1}{6}$ vacancies	defect NaCl	doesn't exist	NaCl structure	$\left(\begin{array}{c}\text{NaCl}\\\text{structure}\end{array}\right)$*	NaCl structure	NaCl structure	PtS structure	wurtzite structure $\left(\begin{array}{c}\text{NaCl at}\\\text{high pressure}\end{array}\right)$
Defect structure			$Ti_{1-\delta}O$ Ti vacancies (intergrowths of $TiO_{1.00}$ and $TiO_{1.25}$ structures)	similar to TiO		$Mn_{1-\delta}O$ Mn vacancies	$Fe_{1-\delta}O$ Fe vacancies in defect clusters	$Co_{1-\delta}O$ Co vacancies	$Ni_{1-\delta}O$ Ni vacancies		$Zn_{1+\delta}O$ interstitial Zn
Conductivity of stoichiometric compound			metallic	metallic <120 K		insulator	(insulator)*	insulator	insulator		insulator
Conductivity of non-stoichiometric compound			metallic	metallic		p-type hopping semiconductor	p-type	p-type	p-type		n-type
Magnetism			diamagnetic	diamagnetic		paramagnetic $\mu = 5.5\mu_B$ (antiferromagnetic when cooled, $T_N = 122 \text{K}$)	paramagnetic (antiferromagnetic when cooled, $T_N = 198 \text{K}$)	antiferromagnetic ($T_N = 292 \text{K}$)	antiferromagnetic (paramagnetic when heated, $T_N = 530 \text{K}$)		

* In practice, *exactly* stoichiometric FeO is never found.

Further along the series, we saw that stoichiometric MnO, FeO, CoO and NiO are *insulators*. This situation is not easily described by band theory because the d orbitals are now too contracted to overlap much—typical band widths are 1 eV—and the overlap is not sufficient to overcome the localising influence of interelectronic repulsions. (It is this localisation of the d electrons on the atoms that gives rise to the magnetic properties of these compounds that are discussed in Section 7.)

Going back now to the non-stoichiometric oxides, in the excess-metal monoxides of type A and type B, we saw that extra electrons have to compensate for the excess metal in the structure. Figure 66 shows that these could be associated either with an anion vacancy or alternatively with metal cations within the structure. Although we have described this association as *reducing* neighbouring cations, this association can be quite weak, and these electrons can be free to move through the lattice: they are not necessarily strongly bound to particular atoms. Thermal energy is often sufficient to make these electrons move, and so conductivity *increases* with temperature.

□   What sort of conductivity do we associate with such behaviour?

■   Semiconductivity. (Metallic conductivity *decreases* with temperature.)

SLC 1   At Second Level, we discussed semiconductivity in terms of band theory. An intrinsic semiconductor has an empty conduction band lying close above the filled valence band. Electrons can be promoted into this conduction band by heating, leaving positive holes in the valence band: the current is carried both by the electrons in the conduction band and by the positive holes in the valence band. Semiconductors such as silicon can also be doped with impurities to enhance their conductivity. For instance, if a small amount of phosphorus is incorporated into the lattice, the extra electrons form impurity levels near the empty conduction band and are easily excited into it. The current is now carried by the electrons in the conduction band and the semiconductor is known as *n-type* (n for negative). Correspondingly, doping with Ga increases the conductivity by creating positive holes in the valence band and such semiconductors are called *p-type* (p for positive).

□   What type of semiconductor would compounds of type A and B produce?

■   n-type, because the conduction is produced by electrons.

Conduction in these non-stoichiometric oxides is not easily described by band theory, for the reasons given earlier for their stoichiometric counterparts—the interelectronic repulsions have localised the electrons on the atoms. So it is easiest to think of the conduction electrons (or holes) localised or trapped at atoms or defects in the crystal rather than delocalised in bands throughout the solid. Conduction then occurs by jumping or *hopping* from one site to another under the influence of an electric field. In a perfect ionic crystal, where all the cations are in the same valence state, this would be an extremely energetic process. However, when two valence states, such as M^{2+} and M^{3+}, are available as in these transition metal non-stoichiometric compounds, the electron jump between them does not take much energy.

Although we cannot develop this theory here, we can note that the conduction in these so-called **hopping semiconductors** can be described by the equations of diffusion theory in much the same way as we did in the last Section for ionic conduction: we find that the mobility of a charge carrier (either an electron or a positive hole), μ, is an *activated process*, and we can write:

$$\mu \propto \exp\left(-E_a/kT\right) \qquad 25$$

where E_a is the activation energy of the hop, and is of the order of 0.1–0.5 eV. The hopping electronic conductivity, σ, is given by the expression

$$\sigma = ne\mu \qquad 26$$

where n is the number of mobile charge carriers and e is the electronic charge. (Notice that these equations are analogous to equations 13 and 8 in Section 5, describing *ionic* mobility, u, and *ionic* conductivity, κ.) The number of mobile carriers, n, depends only on the composition of the crystal, and does not vary with temperature. From equation 26, we can see that, as for ionic conductivity, the hopping electronic conductivity increases with temperature.

In the type C and D monoxides, we have shown the lack of metal as being compensated by oxidation of neighbouring cations to M^{3+}. The M^{3+} ions can be regarded as M^{2+} ions associated with a positive hole. Accordingly, if sufficient energy is available, conduction can be thought to occur via the positive hole hopping to another M^{2+} ion, and the electronic conductivity in these compounds will be p-type. MnO, CoO, NiO and FeO are materials that behave in this way. We saw this behaviour of hopping semiconduction described for NiO in Section 4.5, when it was described in terms of *electron* hopping. Regarding the charge carriers as positive holes is simply a matter of convenience, and the description of a positive hole moving from Ni^{3+} to Ni^{2+} is the same as saying that an electron moves from Ni^{2+} to Ni^{3+}.

Non-stoichiometric materials can be listed that cover the whole range of electrical activity from metal to insulator. Here, we have considered some metallic examples that can be described by band theory (TiO, VO) and others (such as MnO) that are better described as hopping semiconductors. Other cases, such as WO_3 and TiO_2, fall in between these extremes, and a different description again is needed. We also met non-stoichiometric compounds, such as calcia-stabilised zirconia and β-alumina, in Section 5, which are good ionic conductors. Indeed, stabilised zirconia exhibits *both* electronic and ionic conductivity, though, fortunately for its industrial usefulness, the former only occurs at low oxygen pressures. It is thus difficult to make generalisations about this complex behaviour, and each case is best treated individually.

The semiconductor properties of non-stoichiometric compounds are extremely important to the modern electronics industry, which is constantly searching for new and improved materials. Much of their research is directed at extending the composition range and thus the properties of these materials. The composition range of a non-stoichiometric compound is often quite narrow, so to extend it (and thus extend the range of its properties also) the compound is doped with an impurity. To take one example: if we add Li_2O to NiO and then heat to high temperatures in the presence of oxygen, Li^+ ions become incorporated in the lattice and the resulting black material has the formula $Li_xNi_{1-x}O$, where x lies in the range 0 to 0.1. The equation for the reaction (using stoichiometric NiO for simplicity) is given by:

$$\tfrac{1}{2}xLi_2O + (1-x)NiO + \tfrac{1}{4}xO_2 = Li_xNi_{1-x}O \qquad\qquad 27$$

☐ What do you expect to compensate for the presence of the Li^+ ions?

■ Ni^{2+} ions will be oxidised to Ni^{3+} or the equivalent of a high concentration of positive holes located at Ni^{2+} cations.

This process of creating electronic defects is called **valence induction**, and it increases the composition range of NiO tremendously. Indeed, at high Li concentrations, the conductivity approaches that of a metal (although it still exhibits semiconductor behaviour).

SAQ 37 ZnO is a type B (excess metal) material. What do you expect to happen to its electronic properties if it is doped with Ga_2O_3 under reducing conditions?

6.6 Conclusions and summary of Section 6

We have tried to give some idea of the size and complexity of this subject, without letting it become overwhelming. The main point to emerge from our explorations is that the concept of random, isolated point defects does *not* explain the complex structures of non-stoichiometric compounds, but that there are many different ways of defects becoming either ordered or even eliminated.

Now try to summarise the material in this Section, and then compare it with our summary below:

1   Non-stoichiometric compounds are found when impurities are introduced into ionic crystals or when the crystals contain metals with a variable valency.

2   Non-stoichiometric compounds exist over a range of composition; the unit cell size varies smoothly with composition, but the symmetry does not change.

3   Density measurements give information on the type of defect present—either vacancy or interstitial.

4   Non-stoichiometric compounds appear *not* to contain random point defects. Defects can order themselves in various ways—in clusters (FeO, UO_2), throughout the lattice (TiO), or they can be eliminated by the partial collapse of the structure (CS planes).

5   Point defects in non-stoichiometric compounds are accompanied by electronic defects.

7 MAGNETIC SOLIDS

Five basic types of magnetism are found in solids: you have already met two kinds earlier in the Course—*diamagnetism* and *paramagnetism*. Diamagnetism is a property of *all* substances when small magnetic moments are induced in opposition to the applied magnetic field: it is a very weak effect, often swamped by other magnetic properties. Paramagnetism is due to the presence of unpaired electrons on atoms or molecules.

Diamagnetism and paramagnetism differ in the sign of the *magnetic susceptibility*, χ, of the material in question. Recall that in a vacuum

$$B = \mu_0 H \qquad\qquad \textbf{28}$$

where B is the *magnetic flux density*, μ_0 is the *permeability of a vacuum*, and H is the strength of the *magnetic field*. The magnetic flux density changes when a magnetic sample is introduced in place of a vacuum so that

$$B = \mu_0(H + M) \qquad\qquad \textbf{29}$$

where M, the *intensity of magnetisation*, is proportional to H

$$M = \chi H \qquad\qquad \textbf{30}$$

and thus

$$B = \mu_0 H(1 + \chi) \qquad\qquad \textbf{31}$$

χ is negative for diamagnetic materials and positive for paramagnetic materials. Thus, the magnetic flux density in a diamagnetic material is lower than in the exterior vacuum: for a paramagnetic material, it is higher. This is illustrated in Figure 67. Diamagnetic materials are repelled by an inhomogeneous magnetic field; paramagnetic materials are attracted, and this allows measurement of χ by the Gouy method.

Interestingly, superconductors, which you will meet in Section 8, are perfect diamagnets, and the magnetic flux density in them is zero—*no* lines of force can pass through a superconductor.

In paramagnetic solids, such as transition-metal complexes, the set of unpaired electrons in one complex does not interact with the set in another because the complexes are quite far apart. The different sets are therefore oriented randomly with respect to one another, and such compounds are said to be **magnetically dilute compounds**.

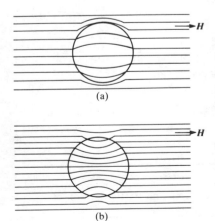

Figure 67 Behaviour of (a) diamagnetic and (b) paramagnetic substances in a magnetic field.

Ferromagnetism occurs in solids when the individual or sets of unpaired electrons are no longer arranged randomly: the spins are *ordered*, so that they are aligned *parallel* to one another throughout a sizeable region known as a magnetic domain (Section 7.3.1). Ferromagnetic substances are found experimentally to have very large positive values of χ (typically in the range 10^{-1} to 10^7), and are strongly attracted by a magnetic field. In contrast to a paramagnet, χ depends on H, so that we cannot give a value to χ for Fe (say) without specifying H. More over, we shall see that χ depends also on the past history of the sample, even for a given H.

Antiferromagnetism is found in solids when the spins interact with each other but where adjacent spins are aligned *antiparallel* and no overall magnetic effect is produced.

In the last form of magnetism that we shall consider—**ferrimagnetism**—the spins are also aligned antiparallel. Here, however, there are either unequal numbers of spins in the two orientations or the antiparallel spins are of differing magnitude leaving a net magnetic moment.

7.1 Effects of temperature

As you saw earlier in the Course, the susceptibility of an ideal paramagnetic substance is inversely proportional to temperature. This is expressed as the *Curie law*.

$$\chi = C/T \qquad\qquad\mathbf{32}$$

and is plotted in Figure 68a. Increasing temperature *increases* the thermal motion of the atoms, thereby destroying the alignment of magnetic moments produced by an external field, and *decreasing* the susceptibility.

The susceptibility of a ferromagnetic solid is found to suddenly rise steeply as the temperature is lowered (Figure 68b). In ferromagnetic solids, the atoms interact with one another in such a way that the spins tend to line up with one another even in the absence of an applied field. However, thermal motion opposes the ordering of the spins, and so as the temperature *rises*, there comes a point at which the thermal motion overcomes the natural spin alignment completely; the spins are randomly oriented with respect to one another and the substance becomes paramagnetic. The temperature at which the ferromagnetic character is lost and the solid becomes paramagnetic is called the **Curie temperature**, T_C, and can be seen as a discontinuity in the plot.

In the antiferromagnetic case (Figure 68c), the interaction between atoms produces an antiparallel alignment of spins. As in the other two cases, the spin arrangement becomes more disordered with increasing temperature until a temperature is reached when the substance becomes paramagnetic: this temperature is known as the **Néel temperature**, T_N.

□ Below the Néel temperature, how would you expect the susceptibility to vary with temperature?

■ As the temperature is lowered, the ordering of the spins antiparallel to one another *increases* and so the susceptibility *decreases*.

The behaviour of magnetic substances is summarised in Table 11.

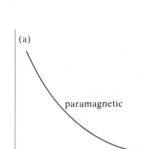
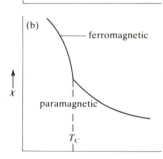
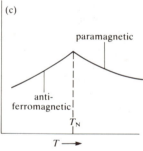

Figure 68 Temperature dependence of the magnetic susceptibility for (a) paramagnetic, (b) ferromagnetic and (c) antiferromagnetic materials.

Table 11 Main types of magnetic behaviour

Type	Sign of χ	Typical χ value (calculated using SI units)	Dependence of χ on H	Change of χ with increasing temperature	Origin
Diamagnetism	−	$-(1\text{–}600) \times 10^{-5}$	independent	none	electron charge
Paramagnetism	+	0–0.1	independent	decreases	spin and orbital motion of electrons on individual atoms
Ferromagnetism	+	$0.1\text{–}10^7$	dependent	decreases	cooperative interaction between magnetic moments of individual atoms
Antiferromagnetism	+	0–0.1	may be dependent	increases	

7.2 Ferromagnetic metals—Fe, Co and Ni

Iron, cobalt and nickel are unusual in being ferromagnetic. No other transition elements possess this property and neither do any main Group metals; only a few of the lanthanides are also ferromagnetic. Why is ferromagnetism confined to such a small number of elements?

Figure 69 shows the crystal structures of Fe, Co and Ni. As you can see they are all different. Other transition metals have structures similar to these (for example, both V and Cr are body-centred cubic), but they do not display ferromagnetism. The crystal structure then does not appear to be the determining factor.

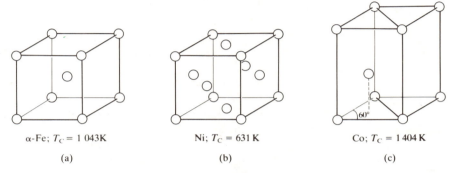

α-Fe; $T_C = 1\,043$K
(a)

Ni; $T_C = 631$ K
(b)

Co; $T_C = 1\,404$ K
(c)

Figure 69 Crystal structures of Fe, Ni and Co: (a) α-Fe, body-centred cubic; (b) Ni, face-centred cubic; (c) Co, hexagonal close-packing.

Iron, cobalt and nickel are all located towards the end of the first transition series, and this turns out to be important. You saw earlier that bonding in metals is usually treated in terms of band theory. For transition metals, the binding comes from the 4s–4p band and the 3d band. Now the 4s–4p band, as in the main Group metals, is wide. The 3d band is narrower and *decreases in width across the transition series*. It overlaps with the 4s–4p band, so that energy-level diagrams for typical transition elements are as shown schematically in Figure 70.

☐ Why does the 3d band decrease in width across the series?

■ The d orbitals contract across the series due to the increased nuclear charge. This leads to less overlap and a narrower band.

Earlier in this Block, we said that electrons fill a band from the lowest energy up with *paired spins*. You may have wondered why the electrons near the Fermi level don't go into higher energy levels with parallel spins. In complexes, where the levels involved were separated by about 10^{-19} J, we saw that it was often favourable to do this. The levels in a band of a typical metal such as sodium are only separated by about 10^{-35} J. Obviously, it would cost little in energy terms to promote one electron in a band.

☐ What are the two competing factors determining whether a complex is high spin or low spin?

■ One is the size of the ligand field splitting, Δ (the energy difference). The other is the difference in electron–electron repulsion between paired and parallel spins.

In a complex, the repulsion between paired spins is greater than between parallel spins because the paired spins are closer together on average. Now, in a metal, the electrons are moving through the whole crystal, so that on average any two spins will be much further apart than in a complex. This applies to both parallel and paired spins, with the result that not only is the interelectronic repulsion less for electrons in a band but also the difference between paired and parallel spins is less. So the advantage in having parallel spins is diminished, and spin pairing is favoured.

In addition, the energy cost in promoting electrons is not as small as it at first sight appears. To observe an appreciable effect in a metal, of the order of n electrons would have to be promoted to higher levels, where n is the number of

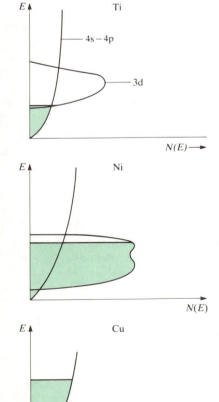

Figure 70 Schematic energy level diagrams for Ti, Ni, and Cu.

atoms in a crystal. Now, some of these would go into the next highest level at a cost of the order of only 10^{-35} J, but in a crystal containing say 0.01 mol of atoms, it is necessary to promote 10^{21} electrons, and there are not 10^{21} levels at this energy. Consequently, higher-energy levels would have to be populated. For main Group metals, and for early transition metals such as Sc, Ti and V, the *net* result is that it is more favourable to pair spins at lower energy than to have electrons with parallel spins in higher energy orbitals.

☐ For iron, cobalt and nickel, the 3d band is narrow. How does this affect the energy needed to promote electrons?

■ The cost in energy in going to levels even at the top of the band is low compared to the energy required to promote electrons to near the top of a wider band such as the 4s–4p.

In addition, since there are five 3d orbitals per atom, there are a very large number of levels in the band and hence a large number at any one energy; that is, the density of states is high. So, there are likely to be more levels available at relatively low energies. Also, the electrons in the 3d band spend more time near the metal nuclei than they would in a metal such as sodium, where free-electron theory is a good approximation, that is the electrons tend to be more localised. This tends to increase the interelectron repulsion and more importantly the difference between parallel and paired spins. Thus, the combination of a larger interelectronic repulsion and smaller promotion energy means that in Fe, Co and Ni the lowest-energy state has large numbers of unpaired electrons. The maximum magnetisation of a piece of iron is that due to $2.2n$ unpaired electrons, where n is the number of atoms in the sample. The ferromagnetism arises from these $2.2n$ electrons *which are all* aligned with their spins parallel* below the Curie point in the *absence* of a magnetic field. This can be contrasted with a solid containing paramagnetic complexes, where the unpaired electrons *on each atom* are aligned parallel to each other but the spins on *different atoms* are aligned randomly in the absence of a magnetic field.

Above the Curie point, the metals are bound only by the 4s–4p band, and the 3d electrons are localised on the metal nuclei. The 3d electrons now produce paramagnetism as in metal complexes.

SAQ 38 A crystal of nickel has a maximum magnetisation corresponding to 0.6 unpaired electrons per nickel atom. If the crystal contains N atoms, how many paired electrons are there in the 4s–4p and 3d bands?

SAQ 39 In addition to iron, cobalt and nickel, the only other elements that exhibit ferromagnetism are some of the lanthanides. Which orbitals would you expect to form narrow bands with a high density of states in this case? Would you expect ferromagnetism to occur at the beginning or end of the series?

SAQ 40 Although manganese is not ferromagnetic, certain alloys of manganese such as Cu_2MnAl are ferromagnetic. The Mn–Mn distances in these alloys are greater than in manganese metal. What effect would this have on the 3d band of manganese? Why would this cause the alloy to be ferromagnetic?

7.3 Properties of ferromagnets

7.3.1 Magnetic domains

By now, some of you will have asked yourselves 'Why isn't every piece of iron a magnet?'. Of course, you know that not all iron is magnetic—I expect that you can remember magnetising a piece of iron at school by stroking it with a permanent magnet. The reason for this is that iron (and other ferromagnetic materials) are composed of **magnetic domains**. Within each domain, the magnetic moments are aligned parallel to give a large magnetisation, but the domains

---

* As you will see in the next Section, the spins are aligned parallel only within domains and not throughout the whole crystal.

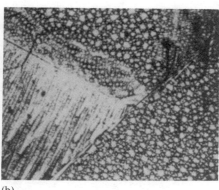

(a)   (b)

Figure 71 (a) Domain patterns for a single crystal of iron containing 3.8 per cent silicon. The white lines show the boundaries between the domains. (b) Magnetic-domain patterns on surface of an individual crystal of iron.

themselves are oriented randomly with respect to each other so there is no *net* magnetisation of the solid. The domains each have a volume of about $10^{-6}\,cm^3$ and contain roughly 10^{15} atoms. Domains can be imaged by electron microscopy as in Figure 71a, or they can be made visible to direct microscopic examination by spreading a very finely divided iron powder on the polished surface of a crystal when the powder lines up in the individual magnetic fields in the domains (Figure 71b).

When a ferromagnetic substance is placed in a magnetic field, its magnetisation increases. This is conventionally shown as a plot of B versus H as in Figure 72. (Ignore the labels on the graph for the moment; we will explain them later.) In a paramagnetic solid, the magnetisation increases because the individual moments line up with the field.

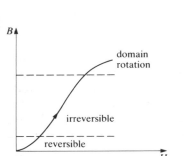

Figure 72 The magnetisation curve plotted as B against H from zero magnetisation up to saturation.

Clearly, in a ferromagnetic solid, this cannot be the case because the individual moments are already aligned. However, they are so aligned only within separate domains. There are then *two* mechanisms by which the magnetisation increases. First, the domains with their magnetic moments aligned with the external field *grow* at the expense of the other domains—this tends to happen at weak fields. Second, in stronger fields, the domains themselves rotate into alignment with the field. These mechanisms are shown diagrammatically in Figure 73, where in (a), we see the four domains with the magnetic moments lined up in four different directions; in (b), a field H has been applied, and the domain aligned with H has grown at the expense of the others; (c) shows the domains beginning to rotate into alignment with the field.

Figure 73 Magnetisation processes according to the domain model: (a) unmagnetised; (b) magnetised by domain growth (boundary displacement); (c) magnetised by domain rotation (spin alignment).

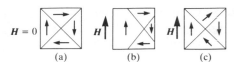

The mechanism of these alignments is not fully understood, but it involves the 'walls' of the domains, which have a finite thickness. Across the width of the wall, the spins *gradually* turn into alignment with the spins of the next domain. It is thought that the domain with moments already aligned with the field grows at the expense of its neighbour by a screw-like motion: the moments undergo a helical twist, which is illustrated in Figure 74 (*overleaf*), effectively moving the wall into the next domain.

Now, we can look back again to Figure 72 and see what the labels mean. At small fields, the domains are able to grow by the mechanism pictured in Figure 74. This is a reversible process, and when the field is removed the ferromagnet loses its magnetisation completely. However, crystals are never perfect, and it takes rather larger fields to make the domains grow through crystal imperfections such as impurities and dislocations. Once a domain boundary has moved through a defect, it finds it difficult to move back again, and so the boundary displacement is permanent. It is only at very high fields that whole domains are able to rotate.

Why do domains occur at all in ferromagnetic solids? From the discussion in the last Section, it would seem favourable for spins to be parallel throughout the

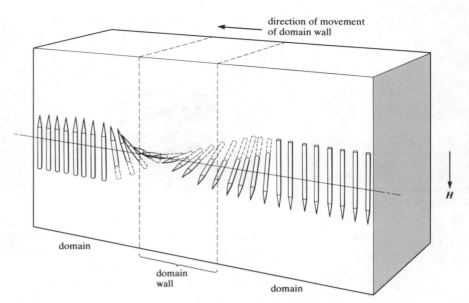

direction of movement
of domain wall

H

domain

domain
wall

domain

Figure 74 The broken lines show the domain wall, which separates two domains with magnetic moments lined up in opposite directions. The moments twist to align with the applied field, *H*, and the wall moves in the direction of the arrow.

whole crystal. The answer lies in the fact that there is another competing interaction. The interelectronic repulsion or exchange mechanism that makes parallel spins more favourable is a *short-range* interaction: its effects fall off very quickly with increasing atomic separation. The competing effect is the **magnetic dipolar interaction** between the spins. This effect is about 1 000 times weaker than the exchange coupling, but is much longer range: the dipolar interaction decreases in proportion to $1/r^3$, where *r* is the distance from the atom in question. The dipolar interaction works in opposition to the exchange coupling and favours neighbouring moments being antiparallel. We can think of the magnetic moment as behaving like a little magnet. As you know, a magnet has two opposite poles, which are usually designated north and south, N and S: opposite poles attract and like poles repel. The same applies to the individual magnetic moments on the atoms. If you have ever handled two bar magnets, you will know that when they are close to each other the stable arrangement is when opposite poles are next to one another. Similarly, individual magnetic moments tend to line up with opposite poles closest to one another and so the spins are *antiparallel*. In summary, there are two competing effects: the interelectronic repulsion is much stronger than the dipolar interaction, but the dipolar interaction is effective over greater distances and therefore over vastly greater numbers of electrons.

If there were only the exchange contribution to worry about, the way to minimise the energy would be for *all* spins to be parallel to one another. However, this happens at the cost of maximising the dipolar energy.

Domains represent a compromise solution to this problem. In each domain, the spins are lined up parallel to one another but in a different direction from that of spins in neighbouring domains. The bigger the domain of parallel spins becomes, the more important the dipolar interaction becomes, until eventually it dominates and it is more stable for the spin to align antiparallel. At this point, a domain wall can form.

At the domain walls, spins are out of line with their nearest neighbours. This is, of course, energetically unfavourable for these atoms. However, the number of atoms near domain walls is small compared with the number deep within the domains, so only a small proportion of the energy in the magnetic field of a fully aligned sample is enough to 'pay' for the creation of domains.

The ease with which a ferromagnet retains or loses its magnetisation is tied up with how easily the domains alter their size and orientation. If we plot the increasing magnetisation of a piece of soft iron against an increasing applied field, we get a curve such as the one drawn in green on Figure 75, which reaches saturation (all spins aligned) at point a. At small fields, the processes producing the magnetisation are reversible, as we described above, and so on removing the field, we would retrace the curve to the origin. If steel is used, some magnetisation remains even after the external field is removed. This is because the favourably

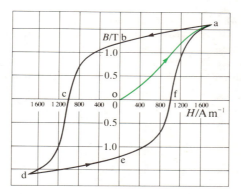

Figure 75 The B–H curve for a typical hard steel.

aligned domains extend themselves by non-reversible processes when they pass through crystal imperfections and then can't return. The B–H plot for a typical steel is shown in Figure 75. As the field is increased, the curve initially follows the green line oa. When the field is reduced below the saturation level and back to zero, the curve follows the line ab. If the field is applied in the opposite direction, at point c the magnetisation is reduced to zero, and at point d saturation in the reverse direction is reached. The magnetisation that remains at point b is termed **remanent magnetisation**, B_r. The complete plot is known as a **hysteresis loop**, as once the iron has been magnetised the magnetisation curve never passes through the origin again. A **permanent magnet** is created by this process. A finite external field called the **coercive force**, opposed to the original field, and equal to the distance oc, is given the symbol H_c, and is a measure of the difficulty of breaking up the alignment of the domains. Materials that retain a high proportion of the original magnetisation, such as the one illustrated in Figure 75, are called **hard magnets**, and those that retain a small or negligible amount are called **soft magnets**.

☐   How will the shape of the hysteresis loop change for a soft magnet?

■   Because the coercive force is very small, the distance oc on the plot will be very small making the loop much 'thinner'.

7.4 Magnetic compounds

7.4.1 Transition-metal oxides

The transition-metal oxides, MnO, FeO, CoO and NiO all have the NaCl structure. They are paramagnetic at high temperature, but become antiferromagnetic on cooling. Taking NiO as our example, the compound has a Néel temperature of 530 K above which it is paramagnetic.

The NaCl structure of NiO is found from X-ray diffraction techniques, which determines the atomic positions. However, when the NiO structure is determined by neutron diffraction techniques, extra magnetic interactions give new peaks in the diffraction pattern; these can be interpreted as giving a **magnetic unit cell** with a unit cell dimension that is *twice* that of the familiar *chemical* unit cell. The structure is shown in Figure 76, where the oxide ions have been omitted for clarity. The close-packed layers of nickel ions lie parallel to the body diagonals of the cubic unit cell—the green shading is only there to help you see these layers more easily. The spins of all the Ni^{2+} ions in a given layer are aligned parallel, but the adjacent layers are aligned antiparallel, giving a net cancellation of the magnetic moments. You will recall from Second Level that lattice points all have identical environments. The normal chemical or crystallographic unit cell in this case must be bounded by identical atoms, whereas the unit cell of the magnetic structure is bounded by identical atoms that *also possess identical spin orientations*. Both unit cells are F-centred in this case. Antiferromagnetic MnO has the same structure, with a Néel temperature of 122 K. The interactions producing these structures are considered in the next Section.

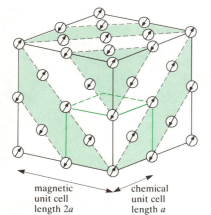

magnetic unit cell length $2a$ chemical unit cell length a

Figure 76 The magnetic unit cell of NiO, with the crystallographic or chemical unit cell indicated in green.

SAQ 41 Cr metal is antiferromagnetic. The chemical structure of Cr is body-centred cubic. The spins can be thought of as arranged antiparallel as shown in Figure 77a. The structure is shown extended in Figure 77b. What type of centring does the magnetic unit cell possess? If the chemical unit cell has a side a, what is the magnetic unit-cell dimension?

Figure 77 Antiferromagnetic ordering in Cr: (a) shows the body-centred (I) crystallographic unit cell, and (b) shows the same structure extended.

Cr; $T_N = 313\,\mathrm{K}$

(a)   (b)

7.4.2 Superexchange

In transition metals, the d orbitals overlap directly. However, in transition-metal *compounds*, the interaction between the metal ions is indirect and involves interaction with the electrons on an intervening ligand—in the case of NiO, this is oxygen. This mechanism is known as **superexchange**.

We have seen that the Ni^{2+} ions are in an octahedral environment.

☐   Sketch the splitting and occupation of the d orbitals in NiO.

■ Ni^{2+} is a d^8 ion with the energy-level diagram shown in Figure 78.

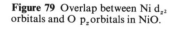

e_g

t_{2g}

$Ni^{2+}\ d^8$

Figure 78 d orbital splitting for Ni^{2+} in an octahedral field.

The Ni^{2+} ions have oxide ions lying midway between them, so we can construct a linear orbital system.

☐   If the Ni—O—Ni direction is labelled z, which orbitals will be in the correct position to overlap?

■   The d_{z^2} on the Ni^{2+} can overlap with the $2p_z$ on the oxide.

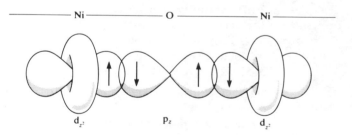

Ni ——————— O ——————— Ni

d_{z^2} p_z d_{z^2}

Figure 79 Overlap between Ni d_{z^2} orbitals and O p_z orbitals in NiO.

The p_z orbital on the oxide ion contains two electrons, while the d_{z^2} orbitals on the nickel ions each contain one unpaired electron. Overlap between the metal d orbitals and the oxygen 2p leads to some covalent mixing—Figure 79. The left-hand $2p_z$ electron will tend to have a spin opposed to that of the left-hand d_{z^2} electron. Bonding to the metal atom on the right-hand side must therefore involve the spin-up electron from oxygen. This is only possible if the right-hand metal has its unpaired electron in the spin-down state. The superexchange mechanism thus predicts that the interaction will be antiferromagnetic, with the two d_{z^2} electrons held in an antiparallel arrangement.

SAQ 42 The compound EuO has the NaCl structure and is paramagnetic above $70\,\mathrm{K}$ but magnetically ordered below it. Its neutron-diffraction pattern at high and low temperatures are identical. What is the nature of the magnetic ordering?

7.4.3 Ferrites

The **ferrites** are a commercially important class of magnetic material, with the general formula AFe_2O_4. One of the ferrites, the mineral magnetite, Fe_3O_4 (where A = iron(II)), is well known as the ancient lodestone or compass: indeed you met this compound earlier in the Course in connection with crystal-field stabilisation energy. You may remember that its crystal structure is known as *inverse spinel*.

The spinels have the general formula AB_2O_4, where A is a divalent ion, A^{2+}, and B is trivalent, B^{3+}. The structure can be thought of as being based on a cubic close-packed array of oxide ions, with A^{2+} ions occupying tetrahedral holes and B^{3+} ions occupying octahedral holes.

☐ What proportion of the tetrahedral and octahedral holes are occupied?

■ In a spinel crystal containing n AB_2O_4 formula units, there are $8n$ tetrahedral holes and $4n$ octahedral holes. There are n A^{2+} ions in tetrahedral holes and $2n$ B^{3+} ions in octahedral holes; accordingly, $\frac{1}{8}$ and $\frac{1}{2}$ of these holes are occupied respectively.

The unit cell is illustrated in Figure 80, and is probably the most complex that you have met so far! We have broken it down into eight octants (of which there are only two kinds): these are shown on the left. The A ions occupy tetrahedral positions in the A-type octants together with the corners and face-centres of the unit cell. The B ions occupy octahedral sites, which take up half the corners of the B-type octants.

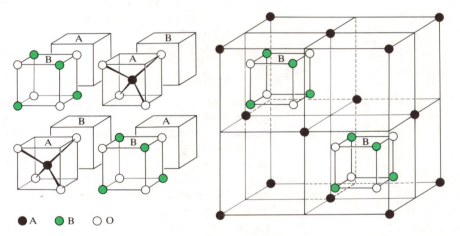

Figure 80 The spinel structure, AB_2O_4.

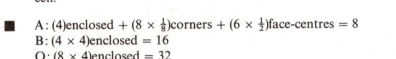
● A   ● B   ○ O

☐ Use Figure 80 to work out the number of formula units AB_2O_4 in the unit cell.

■ A: (4)enclosed + (8 × $\frac{1}{8}$)corners + (6 × $\frac{1}{2}$)face-centres = 8
B: (4 × 4)enclosed = 16
O: (8 × 4)enclosed = 32
giving 8 formula units.

The CFSEs for iron(II) and iron(III) in tetrahedral and octahedral weak-field sites are: octahedral: iron(II) d^6 $\frac{2}{5}\Delta_o$; iron(III) d^5 0, and tetrahedral: iron(II) d^6 $\frac{3}{5}\Delta_t$; iron(III) d^5 0. So iron(II) in octahedral sites will be the most stabilised (Δ_t is only about $\frac{4}{9}\Delta_o$).

We can write the formula of Fe_3O_4 as $Fe^{III}[Fe^{II}Fe^{III}]O_4$ because iron(II) prefers the octahedral sites (which are the B sites in the spinel structure), and so Fe_3O_4 takes the inverse-spinel structure with half of the iron(III) ions also in octahedral sites and the other half in tetrahedral.

The ferrites have interesting magnetic structures, because the ions on tetrahedral sites have magnetic spins that are antiparallel to those of the ions on octahedral sites. This is illustrated in Figure 81. Depending on the nature of the metal ion, A, this can lead to a complete cancellation of magnetic moments, and the ferrite

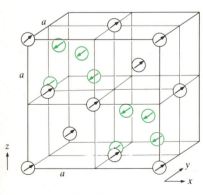

tetrahedral site

octahedral site

Figure 81 Magnetic structure of antiferromagnetic and ferrimagnetic spinels.

becomes *antiferromagnetic*, or to a partial cancellation, in which case the ferrite is *ferrimagnetic*. Let's see what happens in Fe_3O_4: half of the Fe^{3+} ions are on octahedral sites and the other half on tetrahedral.

☐ What will be the net magnetisation due to Fe^{3+}?

■ Zero. There are equal numbers of Fe^{3+} on spin-opposed sites and the magnetic moments will all cancel.

This leaves the iron(II) ions with their spins aligned on octahedral sites producing a net magnetisation. Fe_3O_4 is thus ferrimagnetic.

SAQ 43 $ZnFe_2O_4$ has the inverse-spinel structure at low temperature. What type of magnetism would you expect it to exhibit?

The above discussion only describes the structure of the ferrites, it does not attempt to explain why they have this particular magnetic ordering. The explanation is rather difficult, and lies in the fact that because the A^{2+} and Fe^{3+} ions are both on equivalent octahedral sites, they are close together energetically and electron transfer can take place between them. Less energy is required for this transfer to take place if the moments of the two ions are aligned parallel. This magnetic interaction is called **double exchange**. The antiparallel alignment of the ions on tetrahedral sites with those on octahedral is caused by superexchange.

☐ Would you expect Fe_3O_4 to be a conductor?

■ Yes, it is a good conductor because electron transfer can take place between the iron(II) and iron(III) ions.

The industrial importance of ferrites is discussed in the final part of this Section.

7.5 Application of magnetic properties

The use of magnetic materials is very much a feature of everyday life that we take for granted—from the transformer to the audiocassette. Much of the design of magnets is too technical for us to be concerned with here, and so we have tried to concentrate on applications that show a range of the properties discussed earlier.

7.5.1 Information storage in computers

Some of the ferrites can be used for information storage in computers. For this purpose, it is important that there are two states that can be switched between rapidly and cleanly: these can then be used as 0 and 1 in the binary system.

☐ What two states of magnetism might be useful for this purpose?

■ The two different orientations, N and S, could be used.

☐ If the switching between N and S is sudden and clean, what shape would you expect the hysteresis loop to take?

■ It will be square or rectangular.

Figure 82 shows an idealised loop for such a system (the initial magnetisation curve is not shown).

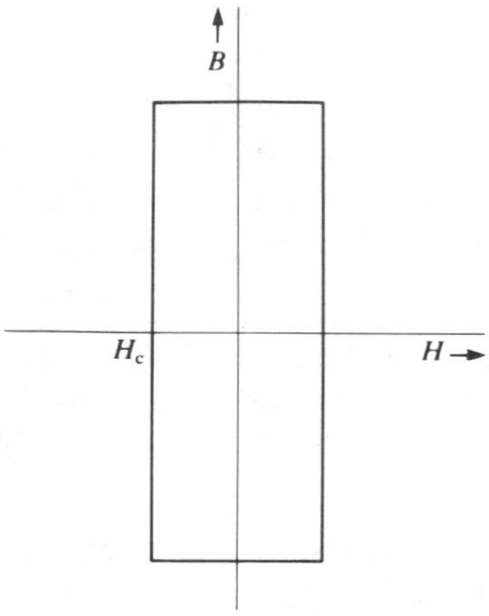

☐ If the sample is magnetised to the saturation point a (Figure 75) what happens when a reverse field is applied?

■ The magnetisation stays constant—as shown by the top horizontal line—until the coercive force, H_c, is reached and the direction of the magnetisation reverses completely.

Switching times are quite fast for some of the ferrites—of the order of 10^{-6} s.

Figure 82 Rectangular hysteresis loop required for information-storage devices.

7.5.2 Tape recording

Sound recordings are made using a polyester tape which is impregnated with a magnetic oxide powder. The oxide most commonly used is γ-Fe_2O_3: chromium dioxide tapes are also popular, and Fe_3O_4 can be used, although recordings are then more difficult to erase.

Briefly, sound recording works in the following way. A sound is due to the movement of molecules in the air—this movement is picked up by the diaphragm in the ear. Similarly, the movement, or sound wave, can make a sensitive diaphragm in a microphone move. This movement is translated into electrical impulses when coupled to a coil which then moves in a magnetic field. The current variation mirrors the original sound wave, and is applied to the magnetic tape via a recording head. The recording head consists of a coil wrapped around an iron core with a gap left where the tape passes across it. The electrical impulses passing through the coil produce a varying magnetic field in the iron, which in turn magnetises the particles on the tape as it passes across the gap. The strength and direction of the magnetisation at a point on the tape are determined by the strength and direction of the magnetic field at the moment the point passes over the head. Thus, the tape becomes a record containing direction (N or S), amplitude (degree of magnetisation) and time (linear distance along the tape). These correspond to the direction, amplitude and time of the electrical signal. When playing the tape back, the whole process works in reverse.

☐　What are the essential properties of the oxide used for coating an audio tape?

◼　It must of course be capable of being magnetised, so is either ferromagnetic or ferrimagnetic. It must also retain its magnetisation—or tapes would be self-erasing!

The oxide coating should also have a fairly high Curie temperature to prevent extremes of temperature demagnetising the tape. γ-Fe_2O_3 fulfills these requirements with a Curie temperature of about 950 K. γ-Fe_2O_3 is made by careful oxidation of Fe_3O_4. (The reaction can be reversed by heating in a vacuum at 520 K (*ca.* 250 °C)). Its structure is based on that of Fe_3O_4. Look back to the unit cell of the spinel structure shown in Figure 80.

☐　In Fe_3O_4, how many iron(III) and iron(II) ions are on octahedral sites, and how many O atoms are there in the unit cell?

◼　The octahedral sites in the diagram are the green circles marked B. There are four in each 'B type' octant making sixteen altogether—all of which are completely contained within the unit cell. There are thus eight of each oxidation state. There are four oxygen atoms in each octant making 32 in all. The remaining eight iron(III) ions are on tetrahedral sites.

When γ-Fe_2O_3 is made from magnetite, the cubic unit cell with the 32 oxide ions is retained, but these are now balanced by iron(III) ions only.

☐　How many iron(III) ions will there be in the unit cell?

◼　$21\frac{1}{3}$ Fe^{3+} ions are now required to balance the oxides.

These $21\frac{1}{3}$ Fe^{3+} ions are found to be statistically distributed over the 24 metal ion positions.

☐　Will γ-Fe_2O_3 be antiferromagnetic or ferrimagnetic?

◼　Ferrimagnetic. Because of the statistical distribution, there will be twice as many Fe^{3+} ions on octahedral sites as there are on tetrahedral. The spins on octahedral and tetrahedral sites are opposed, leaving a net magnetisation due to $\frac{1}{3}$ of the Fe^{3+} ions.

Chromium dioxide, CrO_2, is currently very popular for making audio cassettes. This is because it has a higher magnetisation and so can give better quality reproduction. Its disadvantages are that its Curie temperature is only about 400 K and it is toxic—it is also rather more expensive!

8 SPECIAL TOPICS

8.1 One-dimensional conductors

In recent years, it has been discovered that several classes of compounds and polymers (including organic compounds) exhibit metallic levels of conductivity—sometimes even superconductivity—in one or more dimensions. Such materials are often called **molecular metals**, and indeed we have already met some of these compounds, such as TiO and VO which are three-dimensional conductors, and TiS_2 and ZrCl which are two-dimensional conductors, in earlier Sections. The low-dimensional compounds in particular have raised much interest among physicists and theoreticians, because theoretical models for interaction between atoms and molecules become much simpler when confined to one or two dimensions. However, as well as their intrinsic interest, such compounds have been found to have unusual properties which can be exploited technologically, particularly by the electronics industry.

The substances that conduct in only one or two dimensions are called **low-dimensional metals**. For instance, a one-dimensional metal has metal-like properties along one direction of the crystal and non-metallic properties at right angles to this.

There are many examples in the literature of low-dimensional conductors. Obviously in a Course such as this, we cannot be exhaustive and so have chosen to look closely at two one-dimensional examples in particular.

8.1.1 Platinum chain compounds

The first molecular substances that were found to conduct electricity like metals were the tetracyanoplatinate and bisoxalatoplatinate salts. These were first prepared in the nineteenth century, but it wasn't until the late 1960s that K. Krogmann noticed their unusual conductivities, and these compounds are often named after him. The parent compound, $K_2Pt(CN)_4 . 3H_2O$, is an insulator.

☐ What is the oxidation state of Pt in $K_2Pt(CN)_4 . 3H_2O$?

■ Platinum(II)—the complex ion present is $[Pt(CN)_4]^{2-}$.

When this salt is dissolved in water and oxidised with a little bromine, copper-coloured needles of $[K_2Pt(CN)_4]Br_{0.3} . 3H_2O$ can be crystallised out. This salt is commonly known as KCP (from potassium (**K**) tetracyanoplatinate(II): the presence of the bromine is occasionally indicated, KCP(Br), by some authors.

☐ How has the oxidation state of Pt changed from the parent compound?

■ It has increased: the conventional rules lead to a value of 2.3. As the common oxidation states of platinum are +2 and +4, it is reasonable to infer that some of the platinum(II) atoms will have been oxidised by the bromine to platinum(IV).

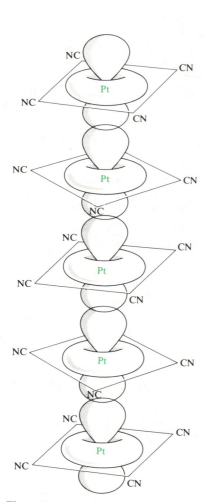

Figure 83 Columnar-stacked structure of $[Pt(CN)_4]^{x-}$ ions.

The KCP crystals contain chains of closely spaced square-planar $[Pt(CN)_4]^{2-}$ ions (Figure 83), and they conduct electricity in the direction of these chains. The conductivity parallel to the chains, $\sigma_\parallel$, is $3 \times 10^4 \Omega^{-1} m^{-1}$. The *partial* oxidation of $K_2Pt(CN)_4$ must be crucial to the conductivity because the platinum(IV) salt, $K_2Pt(CN)_4Br_2$, like the platinum(II) complex, $[K_2Pt(CN)_4]$, does not conduct at all.

Figure 83 shows that the square-planar $[Pt(CN)_4]^{2-}$ ions are stacked parallel to each other but with each alternate ion rotated through 45°.

☐ Why should each complex ion turn through 45° like this?

■ In this way, interatomic repulsions between the cyanide ligands are minimised—the atoms are as far apart as they can get.

The Pt–Pt separation is 289 pm along the chain; this is quite close to the Pt–Pt distance of 277 pm found in Pt metal, and indicates the presence of metallic bonding.

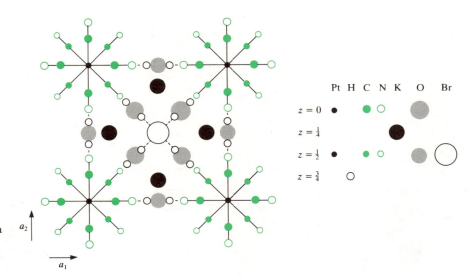

Figure 84 Crystal structure of KCP, showing a projection along the conducting c-axis. We are looking down a Pt chain at each corner of the unit cell: the staggering of the CN ligands can be seen.

Figure 84 shows the packing diagram for KCP. The bromide ions are found in the centre of the unit cell, although only 60 per cent of the available sites are occupied (in a random fashion). Water molecules and potassium ions also occupy the space between the chains.

The ligands (cyanide in this case) surrounding the metal atoms ensure that the interchain separation is large—about 800 pm—so there is no overlap of orbitals at right angles to the stacking direction. These columnar structures (KCP is only one of many) are always found to have ligands composed of atoms from the first short period, typically cyanide and oxalate.

☐ Can you think of a reason why?

■ The smaller atoms allow a sufficiently close approach of the metal atoms within the chain for metallic bonding to take place.

As noted in Section 4, metallic bonding is more probable in the second and third transition series where the d orbitals are more extended. Thus, most one-dimensional metallic complexes are found for platinum(II) and iridium(I) where the d^8 configuration favours the square-planar shape suitable for stacking, and the diffuse 5d orbitals favour metallic bonding. The type of ligand used is also important in the formation of this type of compound: π-acceptor ligands such as CO, CN^-, $[C_2O_4]^{2-}$ are helpful, as they withdraw electron density from the metal atoms and so reduce coulombic repulsions, which could prevent the metal-atom chain from forming. Table 12 lists a selection of these compounds.

Table 12 Metal chain compounds showing one-dimensional conductivity

Compound	Abbreviation	Pt–Pt bond length/pm	Conductivity $\overline{\Omega^{-1}m^{-1}}$
$K_2[Pt(CN)_4]Br_{0.3} \cdot 3H_2O$	KCP(Br)	289	30×10^3
$Rb_2[Pt(CN)_4](FHF)_{0.4}$	RbCP(FHF)	280	230×10^3
$Cs_2[Pt(CN)_4]Cl_{0.3}$	CsCP(Cl)	286	20×10^3
$K_{1.75}[Pt(CN)_4] \cdot \frac{3}{2}H_2O$	K(def)CP	296	$0.5\text{–}10 \times 10^3$
$Rb_{1.67}[Pt(C_2O_4)_2] \cdot \frac{3}{2}H_2O$	Rb–OP	$\begin{cases} 272 \\ 283 \\ 302 \end{cases}$	0.7
$Mg_{0.82}[Pt(C_2O_4)_2] \cdot 6H_2O$	Mg–OP	285	$0.02\text{–}5 \times 10^3$

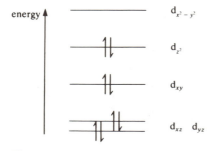

Figure 85 Energy-level diagram for a square-planar d^8 system.

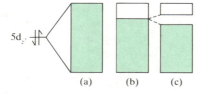

Figure 86 Simple band structure of a one-dimensional metallic complex. (a) filled d band; (b) partially filled band; (c) splitting of a partially filled band by the Peierls distortion.

Now let's see how these compounds conduct electricity. As we have already said $[Pt(CN)_4]^{2-}$ is a square-planar d^8 system.

☐ Sketch the energy-level diagram you would expect to find for the d orbitals and fill in the electrons. What is the highest occupied orbital?

■ You should expect your diagram to look like the one in Figure 85, with d_{z^2} being the highest occupied orbital.

☐ If adjacent Pt atoms are to form a Pt–Pt bond, which orbitals would be in the correct orientation for overlap (assume the z-direction lies along the chain)?

■ The most obvious candidates for overlap are the $5d_{z^2}$ orbitals.

Indeed these are the orbitals illustrated overlapping in Figure 83. Other orbitals, which could contribute to a much lesser extent, are p_z, d_{xz}, d_{yz} and $d_{x^2-y^2}$.

If the Pt atoms in the chain are close enough together, the overlapping $5d_{z^2}$ orbitals form a one-dimensional band throughout the crystal, as shown in Figure 86a.

☐ Would you expect this crystal to conduct?

■ On the face of it, no, because the highest-energy orbitals for $[Pt(CN)_4]^{2-}$ are the d_{z^2} orbitals; these are full, and so, therefore, is the band.

How, then, do the metallic properties arise? When KCP is formed by oxidation with bromine, the electrons needed to form the bromide ions in the complex are ionised from the top of the d_{z^2} band. The new partially filled band (Figure 86b) permits one-dimensional metallic conductivity along the direction of the platinum chains. (The crystals also demonstrate other properties of metals in this direction, such as the reflection of light.)

☐ How many electrons are removed from the top of the band?

■ From the formula for KCP, $[K_2Pt(CN)_4]Br_{0.3} \cdot 3H_2O$, we know that 0.3 electrons are removed from the band, leaving 1.7 electrons per platinum in the $5d_{z^2}$ band instead of 2 as before.

(This, of course, is averaged over the whole crystal structure and does *not* mean that we have fractional electrons!) In KCP, the partial oxidation of the Pt atoms is compensated by the bromide ions.

☐ Can you see from Table 12 how else such compensation can be made?

■ A deficiency of cations also compensates, as in $Rb_{1.67}[Pt(C_2O_4)_2] \cdot \frac{3}{2}H_2O$, for instance.

When one looks closely at the properties of KCP, some interesting observations are made which need further explanation. Firstly, KCP is only metallic in the region of room temperature. As the temperature is lowered, the conductivity is found to drop sharply below 150 K.

☐ Would you expect this behaviour in a metal?

■ No; as you saw in Section 2, conductivity decreases with increasing temperature in metals, and in KCP this is only true in the region of room temperature.

Obviously a band gap forms at lower temperatures causing KCP to behave like a semiconductor. Can we explain this?

The explanation lies with **Peierls' theorem**, which asserts that a one-dimensional metal is always electronically unstable with respect to a non-metallic state—there is always some way of opening an energy gap and creating a semiconductor.

To see how this is implemented it will help to look at some careful X-ray diffraction data on another Pt-chain compound—$Rb_{1.67}[Pt(C_2O_4)_2] \cdot \frac{3}{2}H_2O$— the cation-deficient conductor just mentioned. This crystal contains chains of six Pt atoms, but *the Pt–Pt separations are not all the same*. In fact, three distinct Pt–Pt distances are found—272 pm, 283 pm and 302 pm: these repeat regularly throughout the crystal. In Figure 87, this is illustrated by showing the Pt atom positions by black dots and then adding vertical lines between any two atoms to represent the bond lengths (all the bond distances are exaggerated for clarity). Can you see that the bond lengths vary in a wavelike fashion throughout the crystal? This periodic variation is said to be like a sine wave or *sinusoidal* in character. It is this variation in bond length along the chain that gives rise to the band gap. In this compound, $Rb_{1.67}[Pt(C_2O_4)_2] \cdot \frac{3}{2}H_2O$, it was possible to determine the bond lengths by traditional X-ray methods: the bond-length differences were fairly large and so were measurable, and (more importantly) the sinusoidal variation happened to coincide with the crystal lattice in a simple fashion. Often these conditions do not hold, and this can lead to great difficulties in structure determination.

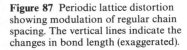

Figure 87 Periodic lattice distortion showing modulation of regular chain spacing. The vertical lines indicate the changes in bond length (exaggerated).

Let's get back to KCP. A (very small) sinusoidal variation of Pt–Pt distances occurs along the Pt chains in KCP. The bonding between Pt atoms is stronger where they are more closely spaced. In these regions, the electrons have lower energy and so tend to concentrate there: this regular build-up of electron density along the chain, coinciding with the shorter Pt–Pt distances, is called a **charge-density wave**. The distortion lowers the energy of the occupied orbitals while raising the energy of the empty ones, and so produces a band gap (Figure 86c). The gap produced is essentially the energy required to move an electron from a more strongly bound region to one where atoms are further apart. So, KCP is an illustration of Peierls' theorem—the distorted semiconductor is more stable than the undistorted chain.

Ah, you say, but why, then, does KCP conduct at all, if it is more stable in its semiconducting state? We know it does conduct at room temperature. The reason is that at these higher temperatures the periodic distortion in the chain is smeared out by the thermal vibrations of atoms and the band gap disappears.

You may already have noticed that there is a very strong similarity here to another theorem that you met earlier in the Course—the Jahn–Teller theorem. In this case, any non-linear molecule in a degenerate electronic state will undergo a distortion so as to remove the degeneracy; and this lowers the electronic energy of the system.

An interesting aside has been pointed out by S. Masuo and colleagues in that KCP possesses an example of each of five different types of bonding within the one unit cell! This is very unusual, if not unique.

☐ Think about the structure of this compound, and list as many types of bonding as you can before turning to an artist's perspective view of the unit cell, Plate 8.1 of S343 *Colour Sheet 2*.

■ The diagram shows O—H and C≡N single and triple *covalent bonds* respectively, picked out in blue. The Pt—C bond (purple) is an example of *coordinate covalent* or *dative bonding*. K^+ and Br^- form *ionic bonds* (green). Pt—Pt *metallic bonding* is present (red) and the water molecules are *hydrogen bonded* to —CN and to Br (yellow).

In conclusion, we can say that many fascinating partially oxidised tetra-cyanoplatinate and bisoxalatoplatinate salts have been made, and no doubt researchers will develop more in the future—perhaps salts that will have even higher conductivities or that retain their conductivities down to lower temperatures. Are the compounds commercially useful at all or just chemical curiosities? Unfortunately, the crystals are rather brittle and tend to lose the water of hydration easily, so there are practical problems in their application. It has been suggested that KCP could be used as a humidity sensor, because $\sigma_\parallel$ is strongly dependent on the surrounding humidity levels. Undoubtedly, though, both chemists and physicists will continue to be interested in their physical behaviour from a theoretical point of view.

8.1.2 Polyacetylene

Most organic polymers such as Polythene (polyethene), PVC (polyvinylchloride) and Perspex (polymethylmethacrylate) are electrical insulators. A polymer that is stable and processable, but that can also conduct electricity would be a major technological breakthrough!

For a polymer to be conducting, some electrons in the backbone must be less strongly localised in the bonds. This is true for unsaturated or *conjugated* structures, which have a skeleton of alternate double and single carbon–carbon bonds. As you will see, the alternation of bonds in organic polymers is another illustration of Peierls' theorem in operation. A well known example of a conjugated long-chain polymer showing such a Peierls' distortion is **polyacetylene**, $(CH)_n$: the precursor for this polymer is ethyne (acetylene), C_2H_2, which has a carbon–carbon triple bond (Figure 88).

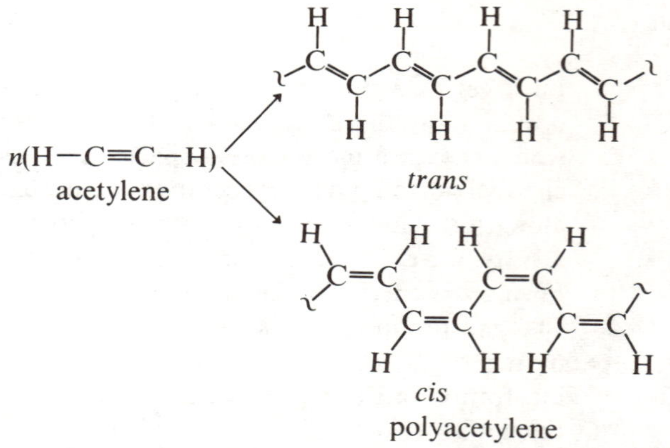

$n(H-C≡C-H)$
acetylene

trans

cis

polyacetylene

Figure 88 Formation of polyacetylene.

The properties of polyacetylene have been much investigated in recent years, but are still under debate because the crystal structure is difficult to determine accurately. Diffraction measurements indicate that there *is* an alternation in C–C bond lengths but only of about 6 pm. This is much less than we would expect for truely alternating single and double bonds (C—C, 154 pm in ethane; C=C, 134 pm in ethene).

☐ Are the electrons completely delocalised in polyacetylene?

■ No: if they were, the C–C bonds would all be identical in length; the diffraction data indicate otherwise.

Now, imagine a form of polyacetylene consisting of a regular evenly spaced chain of carbon–carbon bonds. This should be a good electrical conductor and have a highest energy band that is half-full.

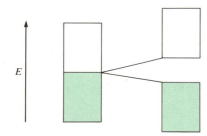

Figure 89 The band gap in polyacetylene produced by the alternation of long and short bonds along the chain.

☐ What does Peierls' theorem say will happen?

■ A periodic distortion will occur that lowers the energy of the occupied orbitals and destabilises the empty ones.

The periodic distortion produces an alternation of long and short bonds along the chain because the electrons are trapped in the bonding orbitals between the closely spaced pairs of atoms. The band gap produced by the distortion (Figure 89) is the energy difference between these bonding orbitals and the corresponding antibonding ones.

☐ From the above discussion would you expect polyacetylene to be a conductor?

■ No, the Peierls distortion has opened up a band gap.

Polyacetylene, in practice, shows modest electrical conductivity, comparable with semiconductors such as silicon: the *cis* form has σ of the order $10^{-7}\,\Omega^{-1}\,m^{-1}$ and the *trans* form, $10^{-3}\,\Omega^{-1}\,m^{-1}$.

In summary, then, the simplest bonding picture for polyacetylene suggests that the π-orbitals overlapping in a regularly spaced chain would give a half-filled band and would be metallic. However, a bond alternation gives a lowering of energy and photoelectron spectroscopy measurements suggest that the band gap is 1.9 eV for the *cis* form and 1.34 eV for the *trans*. In general, conjugated polymers can at best be expected to be semiconductors (although even this property could be useful in a mouldable plastic!).

Apart from the problems of low conductivity, workers encountered other problems in finding a useful conducting polymer: they were only managing to prepare short-chain molecules or amorphous, insoluble, unmeltable powders. However, in 1961 Hatano and co-workers in Tokyo managed to produce thin films of polyacetylene, and ten years later Shirakawa and Ikeda made films of *cis*-polyacetylene which could be converted into the *trans* form. The Japanese achieved this by directing a stream of ethyne gas on to the surface of a Ziegler–Natta catalyst (a mixture of triethyl aluminium and titanium tetrabutoxide). To make a large film, the catalyst solution can be spread in a thin layer over the walls of a reaction vessel (Figure 90), and then ethyne gas allowed to enter. The polyacetylene produced in this way has a smooth shiny surface on one side and a sponge-like structure. It can be converted to the *trans* form (which is thermodynamically more stable) by heating—above 370 K (*ca.* 100 °C) the conversion is quite rapid. After conversion, the smooth side of the film is silvery in appearance, becoming blue when the film is very thin.

The significance of these advances was realised when Shirakawa visited McDiarmid and Heeger in Pennsylvania later in the 1970s. The Americans had been working on smaller conjugated molecules to which they added an electron acceptor in order to make them conducting. It was a natural step to try this approach with polyacetylene. Dopants such as bromine can act as an electron acceptor when added to the polymer: we can think of the structure as $[(CH)^{\delta+}Br_{\delta}^{-}]_{n}$. The

Figure 90 A film of polyacetylene forms on the inner surface of the reaction vessel, after ethyne gas passes over the catalyst solution on the walls. The paper-thin flexible sheet of polyacetylene is then stripped from the walls prior to doping.

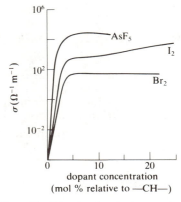

Figure 91 Plot showing the increase in conductivity of polyacetylene on the addition of various dopants.

Figure 92 The conductivity of undoped and doped polyacetylene —$(CH)_n$— compared with values for some of the better-known insulators, semiconductors and metals.

electrons are removed by the bromine and leave holes in the lower part of the π band, rendering conduction possible.

Different dopants can be capable of either oxidising or reducing polyacetylene.

□　What is the role of bromine in the above example?

■　It has oxidised polyacetylene.

Other examples of dopants that can oxidise polyacetylene are I_2, AsF_5 and $HClO_4$: these are all electron acceptors. Polyacetylene can be reduced by electron donors such as the alkali metals to give $[Li_\delta^+(CH^{\delta-})]_n$, for instance. The results are quite dramatic. The conductivity increases from $10^{-3}\,\Omega^{-1}\,m^{-1}$ to as much as $10^5\,\Omega^{-1}\,m^{-1}$ using only small quantities of dopant (Figure 91). This is similar to the conductivity of many metals (see Figure 92).

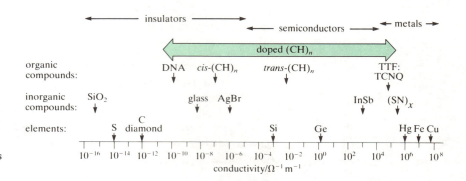

□　How will an electron donor facilitate conductivity?

■　Electrons will enter the top unfilled band making it into a partially filled band.

It is important that the dopant is not only capable of either oxidising or reducing polyacetylene, but that the reaction also provides a counter-ion so that the doped polymer is not left with a net charge. We can illustrate this with a simple example—reduction by lithium. The appropriate half-reactions are:

$$(CH)_n + \delta ne^- = [(CH)^{\delta-}]_n \qquad\qquad 33$$

$$\delta n Li = \delta n Li^+ + \delta n e^- \qquad\qquad 34$$

which can be combined to give

$$\delta n Li + (CH)_n = [Li_\delta^+(CH)^{\delta-}]_n \qquad\qquad 35$$

The highest conductivities for doped polyacetylene—over $10^5\,\Omega^{-1}\,m^{-1}$—have been found using sulphuric or perchloric acid (H_2SO_4 or $HClO_4$) as the dopant. The overall reaction for these acids is quite complicated, as a portion of reactant is used to oxidise polyacetylene while the rest acts as the counter-anion. Using perchloric acid as the example: the half-reaction is

$$ClO_4^- + 8H^+ + 8e^- = Cl^- + 4H_2O \qquad\qquad 36$$

One perchlorate anion is necessary to oxidise polyacetylene, itself being reduced to Cl^-, but a further *eight* perchlorate anions are needed as counter-ions in the polymer, giving the overall reaction:

$$8(CH)_n + 9\delta n HClO_4 = 8[(CH)^{\delta+}(ClO_4)_\delta^-]_n + \delta n HCl + 4\delta n H_2O \qquad\qquad 37$$

The range of conductivity accessible with polyacetylene is amazing—see Figure 92. This is because both donor *and* acceptor molecules can be added to the polymer: if the polymer has been made conducting by the addition of an acceptor molecule, for instance, a donor molecule will interact with it, thus making it ineffective. By adjusting the quantities of donor/acceptor carefully, it should, in theory, be possible to tailor-make the conductivity that you want.

The great interest in these conducting polymers is that they would combine their metallic properties with all the advantages that plastics possess, such as lightness, ease of manufacture and cheapness. Experiments have also been carried out on the use of conducting polymers in solar cells. In a Second Level Course, you met the formation of p–n junctions and their use in photovoltaic cells.

SLC 9

☐   How could polyacetylene be used to form such a cell?

■   By analogy with conventional semiconductors, polyacetylene will be p-type if doped with an electron acceptor and n-type if doped with an electron donor. A p–n junction can be made by bringing two of these films together.

The absorption bands of conductive polymers cover much of the visible spectrum, so sunlight may be converted into electrical energy by photovoltaic cells made of combinations of p- and n-doped polymers. The research shows that it is possible—but the efficiency is low.

Conducting polyacetylene has a variety of possible uses but no practical devices have appeared so far, primarily because of its susceptibility to attack by the oxygen in the atmosphere. The polymer loses its metallic lustre and becomes brittle when exposed to air. However, it seems likely that conducting polymers suitable for commercial exploitation will eventually be found.

We recommend that you now view the sequence on polyacetylene on Videocassette 2, Programme 4, referring to the S343 Audiovision Booklet for guidance.

8.2   Superconductivity

As we write this Course, the whole field of superconductivity is changing dramatically. You may have read in the newspapers or seen on television about the recent (1986) discovery of so-called 'high-temperature' superconductors by Bednorz and Muller. Their findings have been thought to be so important that only a year later the Royal Swedish Academy of Sciences awarded them the Nobel Prize for Physics.

What is all the fuss about? Superconductors have two unique properties that have been said 'could herald a new industrial age' if they could be exploited technologically. First, they have zero electrical resistance and thus carry current with no energy loss: this could revolutionise national grids, for instance, and is already put to good use in the windings of superconducting magnets used in n.m.r. experiments. Second, they expel all magnetic flux from their interior and so are forced out of a magnetic field. These superconductors can float or 'levitate' above a magnetic field and experiments have already been carried out in Japan for a frictionless train floating on magnetic rails, attaining speeds of over $500 \, km \, h^{-1}$ (300 m.p.h.) so far. The possibilities are exciting. However, as always, there is a snag! Until recently, superconductivity has only been found at temperatures close to absolute zero: indeed, until 1986, the highest temperature a superconductor operated at was 23 K, and so they *all* had to be cooled by liquid helium (boiling temperature $\sim 4 \, K$). Helium, as you know, is a fairly rare element, and liquid helium is extremely expensive to produce *and* difficult to handle. This made superconductors simply too expensive to be exploited in any commercial sense. So, for many years, workers have tried to find a superconductor that will operate at the very least at liquid nitrogen temperatures (above 77 K), and eventually they hope for a room-temperature superconductor. (Liquid nitrogen is relatively cheap, because the supply of nitrogen is virtually unlimited.) The events of the past year (1987/8) have brought the operating temperatures of superconductors above the magic liquid-nitrogen temperature, and there is even evidence of superconductivity being observed well above 200 K (though this is not yet for well characterised materials). At the time of writing, no one has yet found a room-temperature superconductor.

As the field is moving so rapidly at present, we will give some of the background to the history of superconductors, their properties and the theory governing their behaviour in the next few pages, but you will find most of the discussion of the

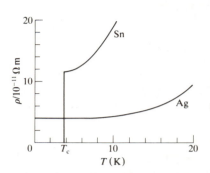

Figure 93 A plot of resistivity ρ versus temperature T, showing the drop to zero at the critical temperature T_c for a superconductor, and the finite resistivity of a normal metal at absolute zero.

new high-temperature superconductors in the S343 *Offprints Folder* where new papers will be added as the subject moves on.

8.2.1 The discovery of superconductors

In 1908, Kammerlingh-Onnes succeeded in liquefying helium. This paved the way for many new experiments to be performed on how materials behaved at low temperatures. For a long time, it had been known from conductivity experiments that the electrical resistance of a metal decreased with temperature. In 1911, Onnes was measuring the variation of the electrical resistance of mercury with temperature when he was amazed to find that at 4.2 K the resistance *suddenly* dropped to zero (Figure 93 shows the same effect for Sn). He called this effect **superconductivity** and the temperature at which it occurs is known as the **super-conducting critical temperature**, T_c. One effect of this zero resistance is that there is no power loss in an electrical circuit made from a superconductor. Once an electric current is established, it shows no discernible decay for as long as experimenters have had the patience to watch (so far about $2\frac{1}{2}$ years)!

More than twenty metallic elements can be made superconducting under suitable conditions (Figure 94), and thousands of alloys.

Figure 94 Superconducting element.

Al	Superconducting
Si	Superconducting under high pressure or in thin films
Li	Metallic but not yet found to be superconducting

| B | Non-metallic elements |
| Fe | Elements with magnetic order |

For more than twenty years, little progress was made in the understanding of superconductors, only more substances exhibiting the effect were found. It was not until 1933 that a new effect (described in the next Section) was observed by Meissner.

8.2.2 The magnetic properties of superconductors

Meissner and Ochsenfeld found that when a superconducting material is cooled below its critical temperature, T_c, it expels all magnetic flux from its interior (Figure 95a).

☐ If the magnetic flux, B, is zero inside a superconductor what is its susceptibility?

■ Since $B = \mu_0 H(1 + \chi)$, then if $B = 0$, χ must be -1. Superconductors are perfect diamagnets.

If a magnetic field is applied subsequently to a superconductor, the magnetic flux is excluded (Figure 95b) and the superconductor repels a magnet. This is shown in the photograph in Figure 96, where a magnet is floating in mid-air above a superconductor.

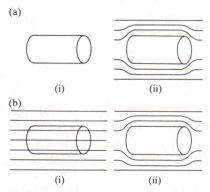

Figure 95 (a) (i) Superconductor with no magnetic field. When a field is applied in (ii) the magnetic flux is excluded. (b) (i) Superconducting substance above the critical temperature, T_c, in a magnetic field. When the temperature drops below the critical temperature (ii), the magnetic flux is expelled from the interior. Both are called Meissner effects.

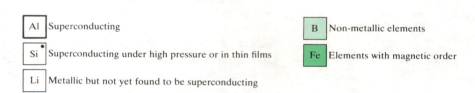

Figure 96 A permanent magnet floating over a superconducting surface.

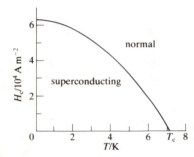

Figure 97 The variation with temperature of the critical field strength, H_c, for lead. Note that H_c is zero when the temperature T equals the critical temperature T_c.

It is also found that the critical temperature, T_c, changes in the presence of a magnetic field. A typical plot of T_c against increasing magnetic field is shown in Figure 97, where you can see that as the applied field increases, the critical temperature drops. It follows that a superconducting material can be made non-superconducting by the application of a large enough magnetic field. The minimum value of the field strength required to bring about this change is called the **critical field strength**, H_c: its value depends on the material in question and on the temperature.

Similarly if the current in the superconductor exceeds a **critical current**, the superconductivity is also destroyed. This is known as the Silsbee effect. The size of the critical current is dependent on the nature and geometry of the particular sample.

8.2.3 The theory of superconductivity

This Section on theory is included partly so that you can see the way in which the subject developed historically, but mainly so that you have some familiarity with the ideas, concepts and terminology used. It will otherwise be difficult for you to follow the review articles in the S343 *Offprints Folder*. The theory of superconductors lies firmly in the realm of theoretical physics and is extremely complex. We do no more here than paint a qualitative picture.

Physicists worked for many years to find a theory that explained superconductivity. To begin with it looked as though the lattice played no part in the superconducting mechanism as X-ray studies showed that neither the spacing nor the symmetry of the lattice changed when the material became superconducting. However, this was shown not to be the case when in 1950 the **isotope effect** was first noticed. For a particular metal, the critical temperature was found to depend on the isotopic mass, M, such that:

$$T_c \propto M^{-1/2} \qquad\qquad \textbf{38}$$

SLC 6   □   Thinking back to a Second Level Course, can you remember the relationship between the vibrational frequency of a diatomic molecule and its mass?

■ $$v = \frac{1}{2\pi}\sqrt{\frac{k}{\mu}} \qquad\qquad \textbf{39}$$

where μ is the reduced mass of the molecule.

Because on isotopic substitution the frequency of a vibration is also found to be proportional to $1/\sqrt{\text{mass}}$, this suggested to physicists that superconductivity was in some way related to the vibrational modes of the lattice and not just to the conduction electrons. You saw in Section 2 that the vibrational modes of a lattice are quantised just as the modes of an isolated molecule are: the quanta of the lattice vibrations are called **phonons**.

This idea was developed further by Frohlich who, in 1950, suggested that there could be a strong phonon/electron interaction in a superconductor that leads to an *attractive* force between two electrons strong enough to overcome the Coulomb repulsion between them. Very simply, the mechanism works like this: as a conduction electron passes through the lattice, it can disturb some of the positively charged ions from their equilibrium positions, pushing them together and thus giving a region of increased positive charge density. As these oscillate back and forth, a second electron passing by this moving region of increased positive charge density is attracted to it. The net effect is that the two electrons have interacted with one another but they have done so using the lattice vibration as an intermediary. Furthermore, the interaction between the electrons is *attractive* because each of the two separate steps involved an attractive Coulomb interaction.

Interestingly, as you saw in Section 2, it is the scattering of conduction electrons by the lattice phonons that produces electrical resistance at room temperature (at low temperatures, scattering by defects predominates).

☐ Would you expect a superconductor to have high or low resistance at room temperature?

■ High—because a superconductor must have strong electron/phonon interactions. Thus the best room-temperature conductors (Ag, Cu) do not superconduct at all!

This is contrary to what we might have expected, but it means that *superconductors do not have low electrical resistance above the critical temperature*.

The final stages of the argument were made by Bardeen, Cooper and Schrieffer who published their theory in 1957. The **BCS theory** (the name is taken from their initials) shows that under certain conditions, the attraction between two conduction electrons due to a succession of phonon interactions can slightly *exceed* the repulsion that they exert directly on one another due to the Coulomb interaction of their like charges. The two electrons are thus weakly bound together forming a so-called **Cooper pair**. It is these Cooper pairs that are responsible for superconductivity.

BCS theory shows that there are several conditions that have to be met for a sufficient number of Cooper pairs to be formed and superconductivity to be achieved. It is beyond the scope of this Course to go into these in any depth; suffice it to say that among these are that the electron/phonon interaction must be strong and that low temperature favours pair formation—hence high-temperature superconductors are not predicted by BCS theory.

Because Cooper pairs are weakly bound, the two members are far apart with a typical separation of 10^6 pm! They are also constantly breaking up and reforming (usually with other partners). There is thus enormous overlap between different pairs and the pairing is a complicated dynamic process. The ground state of a superconductor is a 'collective' state, describing the ordered motion of large numbers of Cooper pairs. When an external electric field is applied, the Cooper pairs move through the lattice under its influence. However, they do so in such a way that the ordering of the pairs is maintained: so the motion of each pair is locked to the motion of all the others, and none of them can be individually scattered by the lattice. This is why the resistance is zero and the system is a superconductor.

SAQ 44 A party of late-night revellers, returning home, decide to take a short cut across a field. Unfortunately, the night is dark and moonless and the field is known to contain deep potholes. Someone has the bright idea that if everyone links arms and advances together across the field, then if one of the company *does* encounter a pothole that one will be lifted clear by dint of collective support! Taking this story as an analogy (not a perfect one of course) with BCS theory, try to identify as many components as possible with corresponding components of BCS theory.

8.2.4 Josephson effects

In 1962, B. D. Josephson predicted that if two superconducting metals were placed next to each other separated only by a thin insulating layer (such as their surface oxide coating) then a current would flow in the absence of any applied voltage. This effect is indeed observed because if the barrier is not too thick then electron pairs can cross the junction from one superconductor to the other without dissociating. This is known as the **d.c. Josephson effect**. He further predicted that the application of a small voltage to such a junction would produce a small alternating current—the **a.c. Josephson effect**. These two properties are of great interest to the electronics and computing industries where they could be exploited for fast-switching purposes.

8.2.5 The search for a high temperature superconductor

By 1973, the highest temperature found for the onset of superconductivity was 23.3 K: this was for a compound of niobium and germanium, Nb_3Ge, and here it stuck until 1986, when Georg Bednorz and Alex Muller reported their findings. As their Nobel Prize citation states 'Last year, 1986, Bednorz and Muller reported finding superconductivity in an oxide material at a temperature 12 °C higher than previously known'. The brilliance and insight that they brought to the subject was to pursue the research into different materials—metallic oxides—for superconductivity. The compound that prompted their initial paper has been shown to be $La_{2-x}Ba_xCuO_4$, where $x = 0.2$, with a structure based on that of K_2NiF_4, a layered perovskite: they observed the onset of superconductivity at 35 K. The structure is said to be layered, as it can be thought of as layers of perovskite-type structure sandwiched between NaCl-type layers. These initial findings are discussed in the article in the S343 *Offprints Folder* entitled 'Record High-Temperature Superconductors Claimed', which is a short review from the journal *Science* by A. L. Robinson, labelled Block 8 Paper A. *Read it now.*

These events were quickly overtaken when physicists realised that it might be possible to raise the critical temperature even higher by substitution with different metals. Using this technique, it was the group at Houston, Texas under the leadership of Paul Chu, that finally broke through the liquid-nitrogen temperature barrier with the superconductor that is now known as '1-2-3': this superconductor has replaced lanthanum with yttrium and has the formula, $YBa_2Cu_3O_{7-\delta}$. The onset of superconductivity for this compound occurs at 93 K!

Theoretical physicists have been working hard to try and explain the phenomenon of high-temperature superconductivity with no clear-cut consensus at present. BCS theory does not seem to be applicable, but there does seem to be conclusive evidence that pairs of electrons are still responsible for the superconductivity: the mechanism of how they are formed is a matter of debate.

Superconductivity is currently a very active research area. There have already been four Nobel prizes awarded to people working in the subject: to K-Onnes for the initial discovery; to Bardeen, Cooper and Schrieffer for BCS theory; to Josephson; and most recently to Bednorz and Muller. Who knows what the future will bring? We can only await the results with eager anticipation.

Now read the other Block 8 offprints in the S343 Offprints Folder.

SAQ 45 The perovskite structure is named after the mineral $CaTiO_3$. Some fluorides, ABF_3, sulphides ABS_3 and many oxides adopt it. The perovskite A-type unit cell is shown in Figure 98, so called because it has an A-type atom at the centre. (a) Draw a packing diagram of this unit cell, determine the number of ABO_3 formula units, and describe the coordination geometry around each type of atom. (b) Redraw the structure so that the unit cell has A atoms at each corner. Which atom is at the centre of the cell now? Draw a packing diagram for this new cell.

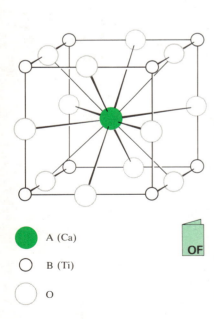

A (Ca)

B (Ti)

O

Figure 98 The perovskite structure for compounds ABO_3, such as $CaTiO_3$.

80

SAQ 46 The unit cell for the K_2NiF_4 structure as adopted by the superconducting compound $La_{2-x}Ba_xCuO_4$ can be divided into three sections of equal lengths along the c direction in Figure 99. What is the structure of the central section? The complete unit cell is made by 'topping and tailing' this central section with two A-type perovskite unit cells which have their bottom layer and top layer missing, respectively. What is the coordination around K/La in this structure? How has this changed from the coordination of an A-type atom in perovskite? Draw packing diagrams for layers at $\frac{1}{6}$ and $\frac{1}{3}$ in c.

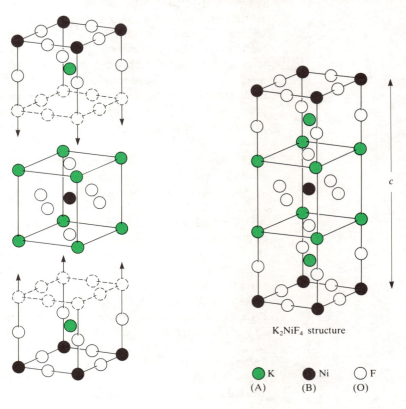

Figure 99 The K_2NiF_4 structure built up from perovskite-type (ABO_3) unit cells.

K_2NiF_4 structure

○K (A)   ●Ni (B)   ○F (O)

8.3 Zeolites

Those of you who have studied the Third Level Course S342 will know of the importance of **zeolites** in catalysis and as molecular sieves and in ion exchange materials. The structure of zeolites and their industrial importance is discussed in the fifth sequence on Videocassette 1 and the associated notes.

We recommend that you now view the sequence on zeolite chemistry on Videocassette 1 and the sequence on superconductors on Videocassette 2, Programme 4, referring to the S343 Audiovision Booklet for guidance.

9 CONCLUDING REMARKS

In conclusion, we can see that solid-state chemistry covers a vast variety of topics; we have done little more here than scratch at the surface of each of them. It should however, have given you a sense of how important it is to understand the detailed structures of solids both from a theoretical point of view and also for the purposes of new technology.

We can do no better than to end the Block by quoting some remarks of Peter Day as he concluded a review on low-dimensional solids in 1983; although the science has moved on, the sentiments are very relevant.

'Layer compounds with exchangeable intercalated cations have certainly found application as battery cathodes but so far as I know, none of the other substances mentioned here has yet reached the market place. Is that important though? I think not. Finding materials whose properties challenge theory is a proper function for academic chemists, and the surprises we get extend the horizons of the possible. Where it will lead us no-one can tell.

Materials with unusual properties may generate applications that did not exist before. Will a high temperature superconductor remain a mirage? Who knows? But one thing must be sure: when the search for such properties becomes an affair for chemists the world of substances to be scrutinised is dramatically enlarged. I commend this thought to the chemical community.'

OBJECTIVE FOR BLOCK 8

Now that you have completed Block 8, you should be able to recognise valid definitions of, and use in correct context, the terms, concepts and principles in Table A.

Table A List of scientific terms, concepts and principles used in Block 8

Term	Page No.
a.c. Josephson effect	79
anti-Beevers-Ross (aBR) site	40
antiferromagnetic	58
attempt frequency	34
β-alumina	37
BCS theory	78
Beevers-Ross (BR) site	40
charge-density wave	71
coercive force, H_c	63
concentration meter	37
conduction path	36
conduction planes	39
Cooper pair	78
critical current	77
critical field strength, H_c	77
crystallographic shear (CS)	50
crystallographic shear planes	50
Curie temperature, T_C	58
d.c. Josephson effect	79
defect cluster	45
density of states	11
doping	33

Your understanding of these terms, concepts and principles is tested in the SAQs within the Block.

SAQ ANSWERS AND COMMENTS

SAQ 1   (a) Figure 100 shows the NaCl packing diagram. (b) The close-packed layers lie parallel to the body-diagonal of the cube. (c) The unit cell contains four formula units of NaCl. This is arrived at as follows: the 12 Na^+ ions at the mid-points of the cell edges are each shared by four cells, and there is 1 Na^+ ion at the centre, making 4 Na^+ in all $[(12 \times \frac{1}{4}) + 1] = 4$. The 8 Cl^- ions at the corners are each shared by eight cells, and the 6 Cl^- at the face centres by two cells, making 4 Cl^- in all $[(8 \times \frac{1}{8}) + (6 \times \frac{1}{2})] = 4$. (d) Both Na^+ and Cl^- ions are octahedrally coordinated by 6 ions of the opposite charge.

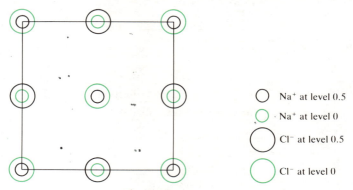

○ Na^+ at level 0.5

◎ Na^+ at level 0

◯ Cl^- at level 0.5

◯ Cl^- at level 0

Figure 100 Packing diagram for NaCl.

SAQ 2   The two lowest levels are $3h^2/8mL^2$ and $6h^2/8mL^2$, the difference between them being $3h^2/8mL^2$. Equation 1 gives the lowest energy level when all three quantum numbers are equal to one, and the next lowest when two quantum numbers are equal to one and the other is equal to two.

SAQ 3 2.01×10^{-34} J or 1.25×10^{-15} eV. The energy difference is $3h^2/8mL^2$ and,

$$\frac{3h^2}{8mL^2} = \frac{3(6.626 \times 10^{-34}\,\text{J s})^2}{8(9.109 \times 10^{-31}\,\text{kg})(3 \times 10^{-2}\,\text{m})^2} = 2.01 \times 10^{-34}\,\text{J}$$

SAQ 4   There are twelve degenerate electron states. For the 3, 2, 1 example, there are six (n_1, n_2, n_3) permutations: (3, 2, 1), (3, 1, 2), (1, 3, 2), (1, 2, 3), (2, 1, 3) and (2, 3, 1). Each permutation can contain two electrons. All twelve states are degenerate because, from equation 1, all have identical values of $(n_1^2 + n_2^2 + n_3^2)$ and, therefore, of E.

SAQ 5   48 states; The appropriate integer sets are 6, 1, 1 and 5, 3, 2 $(n_1^2 + n_2^2 + n_3^2) = 38$; 6, 2, 1 and 4, 4, 3 $(n_1^2 + n_2^2 + n_3^2) = 41$; 5, 4, 1 $(n_1^2 + n_2^2 + n_3^2) = 42$. The sets (5, 3, 2), (6, 2, 1), and (5, 4, 1) contribute 12 states each, the sets (6, 1, 1) and (4, 4, 3) contribute 6 states each.

SAQ 6 $2.04 \times 10^6\,\text{m s}^{-1}$; equating the kinetic energy of an electron ($m = 9.109 \times 10^{-31}\,\text{kg}$) in the Fermi level to the Fermi energy:

$$\tfrac{1}{2}mv^2 = E_F = 11.8\,\text{eV} = 1.89 \times 10^{-18}\,\text{J}$$

$$v = \left[\frac{2 \times 1.89 \times 10^{-18}\,\text{J}}{9.109 \times 10^{-31}\,\text{kg}}\right]^{1/2}$$

$$= 2.04 \times 10^6\,\text{m s}^{-1}$$

SAQ 7   Each chlorine atom has three 3p orbitals, so the lower band is formed from $3n$ atomic orbitals. It therefore contains $3n$ molecular orbitals, and can accommodate $6n$ electrons. As we have classified the Cl 3s electrons as inner, non-bonding electrons, there are six electrons to be accommodated per NaCl formula unit: one from Na 3s and five from Cl 3p. This amounts to $6n$ valence electrons in all, which therefore exactly fill the lower band. Because the Cl 3p valence band is full, and the Na 3s conduction band is empty, and there is a large band gap between the two levels, NaCl is an insulator.

SAQ 8   Figure 19 suggests that the gap between the valence and conduction bands will increase as the difference in the energies of the Na 3s and halogen np atomic orbitals increases. As the ionisation energies of the halogen and sodium atoms run in the order $F > Cl > Br > I > Na$, this difference follows the sequence NaF > NaCl > NaBr > NaI: the correlation with the band gap does indeed exist. If the sodium halides were coloured, there would have to be an electronic transition in the energy range 1.8–3.1 eV. But the transition of *lowest possible energy* is that from the top of the filled valence band to the bottom of the empty conduction band. This is the band gap. So all transitions in the sodium halides have energies in excess of 6 eV, visible light cannot be absorbed, and the compounds are colourless.

SAQ 9 The conductivity should increase. The formulation $Li^+[TiS_2]^-$ suggests that Figure 23c should be applied to the anion $[TiS_2]^-$ in which, compared with TiS_2, there is one extra electron per TiS_2 unit. These extra electrons will be added to the conduction band, raising the Fermi energy well above the area of band overlap where the density of states is low, to a region where there are many empty states immediately above the Fermi level. This predicted increase in conductivity is observed.

SAQ 10 The lower band formed from O 2p orbitals contains six electron states per TiO unit. Four of these are occupied by the 2p electrons of the oxygen atom, and the remaining two by two of the four outer electrons of titanium. This leaves two electrons to be placed in the $d(t_{2g})$ band, which, because it is derived from the three $d(t_{2g})$ orbitals contains six electron states per TiO. Thus the valence band is 100 per cent occupied and the $d(t_{2g})$ conduction band is 33 per cent occupied. The partially filled conduction band accounts for the metallic conductivity.

SAQ 11 One-sixth of the titaniums and one-sixth of the oxygens. From every third plane of titaniums and every third plane of oxygens, one-half of the atoms are removed. Thus the fraction lost is $\frac{1}{3} \times \frac{1}{2} = \frac{1}{6}$. In the context of the NaCl-type structure, the formula of TiO is $Ti_{5/6}O_{5/6}$.

SAQ 12 Ti_2O_3 and ReO_3 should be metallic; WO_3 should be an insulator. We use the band structure diagram of Figure 23a. In Ti_2O_3, ReO_3 and WO_3, there are 18 electron states per unit in the valence band, 12 of which are occupied by the 2p electrons of the oxygen atoms. Tungsten and rhenium have 6 and 7 outer electrons respectively, so in WO_3 the valence band is full and the $d(t_{2g})$ conduction band is empty. In ReO_3, the six electron states per unit in the $d(t_{2g})$ band are one-sixth or 16.7 per cent occupied. In Ti_2O_3, there are eight outer titanium electrons per Ti_2O_3 unit, six of which complete the valence band; the remaining two again give one-sixth or 16.7 per cent occupancy of the 12 electron states per Ti_2O_3 unit in the $d(t_{2g})$ conduction band.

SAQ 13 In the NaCl unit cell of Figure 5, there are 4 formula units. In Figure 29, there are 3 formula units: one oxygen has been removed from the centre of the unit cell, and the effect of eliminating the eight niobiums at the corners is to remove one niobium in all. Thus one-quarter of the niobiums and oxygens have been removed in moving from Figure 5 to Figure 29.

SAQ 14 See Figure 101: In stage A, which represents the NiO insulator, there are two parallel spins in the e_g orbitals. In stage B, these spins are paired up in the same orbital, incurring a pairing energy, P. In stage C, which represents a hypothetical metallic compound, the orbital energy of the e_g electrons is lowered when they half-fill a $d(e_g)$ conduction band. As NiO is an insulator, stage A is a better description than stage C. This shows that the lowering of orbital energy caused by band formation is not sufficient to compensate for the pairing energy, P. To change this, the band width would have to be increased.

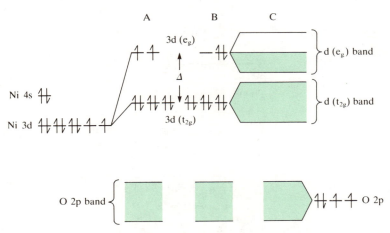

Figure 101 A simplified band structure for a hypothetical metallic form of NiO; the stages A, B and C correspond with those in Figure 30.

SAQ 15 For layer 13; the repeat patterns for position and type of atom respectively are:

A B C A B C A B C A B C A
Cl Zr Zr Cl Cl Zr Zr Cl Cl Zr Zr Cl Cl

Only at layer 13 does the (A, Cl) pairing next occur. Layers 7 and 10 are exactly super-posed on layers 1 and 4, but they contain zirconium and not chlorine atoms.

SAQ 16 Ti_5Te_4; see Figure 102. When the clusters of Figure 33 are joined through opposite corners of the metal octahedra, each cluster unit then contains only five titanium atoms, because it shares half of the two atoms at the opposite linking corners with adjacent clusters. Likewise every tellurium atom in the parent Ti_6Te_8 cluster is shared with an adjacent cluster, so its tellurium atom count is only four.

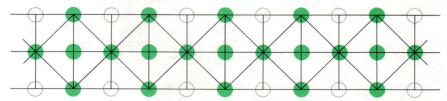

Figure 102 When clusters of the type in Figure 33 with a formula Ti_6Te_8 are joined through opposite corners of the metal octahedra to form a chain, the compound Ti_5Te_4 is the result.

SAQ 17 (a) This is seen most readily by concentrating on the anion at the centre of the unit cell in Figure 39c. This is surrounded by four cations at a distance of $0.43a$ and six anions at a distance of $0.5a$.

(b) Now the nearest neighbours are the eight anions at the corners of the primitive cube (Figure 39d)—at a distance of $0.43a$. The next-nearest neighbours are six cations at a distance of $0.5a$—the cations at the face-centres in Figure 39a.

SAQ 18 From the answer to SAQ 17, for a *normal anion site*, $r = 0.43 \times 537 \times 10^{-12}$ m for inter-action with 4 cations ($Z = +2$), and $r = 0.5 \times 537 \times 10^{-12}$ m for interaction with 6 anions ($Z = -1$). Inserting these values in equation 9 gives:

$$\text{potential energy} = -\left[\frac{2.31 \times 10^{-28}\,\text{J m}}{537 \times 10^{-12}\,\text{m}}\right]\left[\frac{4 \times 2}{0.43} + \frac{6 \times (-1)}{0.5}\right]$$

$$= -(4.302 \times 10^{-19}\,\text{J})(6.605)$$

$$= -2.84 \times 10^{-18}\,\text{J}$$

Similarly, for the *interstitial site*,

$$\text{potential energy} = -(4.302 \times 10^{-19}\,\text{J})\left[\frac{8 \times (-1)}{0.43} + \frac{6 \times 2}{0.5}\right]$$

$$= -(4.302 \times 10^{-19}\,\text{J})(5.395)$$

$$= -2.32 \times 10^{-18}\,\text{J}$$

The energy of defect formation is the energy change for the process: normal site → intersti-tial site, so is given by the difference,

$$(-2.32 \times 10^{-18}\,\text{J}) - (-2.84 \times 10^{-18}\,\text{J}) = 5.2 \times 10^{-19}\,\text{J}$$

SAQ 19 **Table 13** Values of c_S/n

T	$\Delta H_S = 5 \times 10^{-19}$ J	$\Delta H_S = 10^{-19}$ J
300 K	6.12×10^{-27}	5.72×10^{-6}
1 000 K	1.37×10^{-8}	2.67×10^{-2}

The answers are collected in Table 13. A sample calculation, with $\Delta H_S = 5 \times 10^{-19}$ J and $T = 300$ K is as follows:

$$\frac{\Delta H_S}{2kT} = \frac{5 \times 10^{-19}\,\text{J}}{2 \times (1.380\,662 \times 10^{-23}\,\text{J K}^{-1}) \times (300\,\text{K})}$$

$$= 60.3575$$

Then $c_S/n = \mathrm{e}^{-60.3565}$

$$= 6.12 \times 10^{-27}$$

(How you work out e^x will depend on your calculator, but it will undoubtedly involve entering the index (60.3575), changing the sign (to -60.3575) and then: *either* pressing the button labelled $\boxed{e^x}$; *or* pressing the function button $\boxed{\text{inverse}}$ (or $\boxed{\text{inv}}$) followed by $\boxed{\text{ln}}$.)

SAQ 20   A pure oxide with the fluorite structure must be of formula type MO_2—and hence comprises more highly charged cations (M^{4+}) and anions (O^{2-}) than a fluoride. This will tend to strengthen the coulombic interactions within the crystal, and hence increase the energy of defect formation (cf. SAQ 18)—as confirmed by the experimental values in Table 5—and reduce the concentration of defects at a given temperature (cf. SAQ 19).

SAQ 21   (a) Little effect. (b) Charge compensation requires that each Ca^{2+} substitutes for *two* Na^+ ions, thereby creating vacancies on the cation sublattice: since Na^+ is the major charge carrier, the conductivity should increase. (c) Little effect. (d) This time, substitution on the anion sublattice (one O^{2-} ion for two Cl^-) creates anion vacancies: since some current is carried by Cl^- ions, this should increase the conductivity a little (though less than in (b) for comparable additions).

Notice that the substitution mechanism leads to *Schottky-like* defects, with vacancies on one or other sublattice. Alternative mechanisms are possible that lead to Frenkel-like defects (i.e. involving interstitial species): you will learn more about this later in the Block.

SAQ 22   Points 1 and 3 are three-coordinate (or trigonal) positions lying on the triangular faces shared by an adjacent octahedron and tetrahedron. Point 2 is a four-coordinate position at the centre of the tetrahedron. Point 4 is a two-coordinate position on the 'edge' shared between two octahedra.

SAQ 23   The lower (and apparently more direct) route actually takes the cation through a highly crowded two-coordinate position (4), that is likely to be very different in energy from its regular octahedral site. By contrast, in tracing out the upper route, the coordination of the cation changes less sharply (from 6 to 3 to 4 to 3, and back to 6)—so this is likely to be the preferred, lower-energy pathway.

SAQ 24   Your sketch should look something like Figure 103. The jump takes the anion from a tetrahedral site, through a trigonal position on the shared triangular face, and hence to the octahedral site at the body centre.

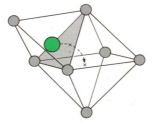

Figure 103

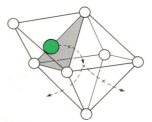

Figure 104

SAQ 25   The main points are as follows:

1   For γ-AgI, the unit cell contains $4\,Ag^+$ and $4\,I^-$ ($8 \times \frac{1}{8}$ at the corners, plus $6 \times \frac{1}{2}$ on the face-centres). Although the unit cell dimensions are different, this does suggest a less open structure than α-AgI.

2   In both structures, Ag^+ ions occupy tetrahedral sites. However, γ-AgI contains an excess of just one available, and equivalent, site per Ag^+, compared with the five in α-AgI.

3   Both structures contain vacant octahedral and trigonal sites. In γ-AgI, the octahedral sites lie at the body-centre, and also at the mid-point of each edge (as in α-AgI). The trigonal sites lie on the triangular faces joining each pair of adjacent tetrahedra and octahedra (as in Figure 103).

4   Like all face-centred cubic structures, γ-AgI can be thought of as alternate octahedra and tetrahedra. Thus, the jump of Ag^+ from one tetrahedral site to another vacant one cannot be as direct as is possible in α-AgI. One route, via an octahedral site, is sketched in Figure 104. Others are possible: for example, the cation could stay in the top half of our octahedron, passing out through a triangular face adjacent to the one it entered through. *All* such pathways will have higher activation energies than the direct route in α-AgI.

5   Both structures obviously contain the same monovalent, polarisable ions.

Points 1, 2 and 4 will contribute to the lower conductivity of γ-AgI (as will the simple fact that it only exists at lower temperatures!).

SAQ 26 I^-, S^{2-} (and heavier members of Group VI) are all 'soft', polarisable anions.

SAQ 27 O^{2-} is a divalent ion, which will increase the strength of its interactions with surrounding ions (compared with F^-). The relative hardness of O^{2-} will do little to mitigate this effect. Experimental data confirm that the activation energies for anionic conduction in oxides are usually higher than for fluorides.

SAQ 28 From the electrode reactions, $n = 4$, and since $a = p/p^\ominus$ for a gas, the Nernst equation for the cell reaction is:

$$E = E^\ominus - (2.303RT/4F) \log (p''/p')$$

In this case, $E^\ominus$ is the *standard* potential of the cell reaction—that is, the value of E when the oxygen on *both* sides of the electrolyte (Figure 49) is at the standard value $p^\ominus \approx 10^5$ Pa (1 atm). Clearly, under these circumstances, there is *no* net tendency for oxygen to move from left to right (or right to left) and no voltage delivered by the cell. In other words, $E^\ominus = 0$, and the expression above becomes:

$$E = -(2.303RT/4F) \log (p''/p')$$

$$= (2.303RT/4F) \log (p'/p'')$$

SAQ 29 In undoped β-alumina, the excess Na^+ ions are balanced by additional oxygens in the conduction plane. Beyond a certain point, it seems likely that these will begin to block the motion of the Na^+ ions—as observed. By contrast, doping with Mg^{2+} suggests an alternative charge compensation mechanism—simple substitution for Al^{3+} ions in the spinel-like blocks (as in the mineral spinel itself) allows extra Na^+ ions into the conduction planes without the need for oxygen interstitials. This is what happens in practice.

SAQ 30 One possibility is that the sodiums diffuse into the crystal as interstitial atoms. An alternative (which in fact happens) is that the sodium atoms occupy cation sites, thus creating the equivalent number of anion vacancies. They subsequently ionise to form a sodium cation with an electron trapped at the anion vacancy. Such trapped electrons are known as **F-centres** and give rise to beautiful colours in crystals where they occur, such as Blue John, the form of fluorite found in Derbyshire.

SAQ 31 Unit cell volume is $(428.2 \text{ pm})^3 = 7.851\,3 \times 10^{-29} \text{ m}^3$.

Mass of contents for *iron vacancies*:

$$[(4 \times 55.85 \times 0.910) + (4 \times 16.00)]/(N_A \times 10^3) \text{ kg}$$

giving a density of $5.653\,4 \times 10^3 \text{ kg m}^{-3}$.

Mass of contents for *oxygen interstitial*:

$$[(4 \times 55.85) + (4 \times 16.00 \times 1/0.910)]/(N_A \times 10^3) \text{ kg}$$

giving a density of $6.212\,6 \times 10^3 \text{ kg m}^{-3}$.

Comparing these theoretical values with the experimental value, we again see that the evidence supports an iron vacancy model.

SAQ 32 From Table 9, we see that as the Fe : O ratio decreases the unit cell volume also decreases; this is the trend we would expect to see as more vacancies are introduced. If the interstitial model were correct, as the Fe : O ratio decreases, the number of interstitial oxygens rises and we would expect to see a slight *increase* in lattice parameter.

SAQ 33 (a) The central section has *two* Fe^{3+} ions in tetrahedral sites and *seven* vacancies, so it is known as a 7 : 2 cluster. (b) There are 32 oxide anions. There are *two* Fe^{3+} ions and *six* Fe_{oct} ions enclosed within the cluster. The outer layer will be the same as the Koch–Cohen cluster with $(8 \times \frac{1}{8}) = 1$ Fe_{oct} at the corners; $(12 \times \frac{1}{4}) = 3$ Fe_{oct} at the mid-points of the edges; and $(30 \times \frac{1}{2}) = 15$ Fe_{oct} on the faces; this makes 27 Fe ions in total, so the formula would be $Fe_{27}O_{32}$. (c) The 32 oxide ions provide 64 negative charges to be balanced. The two tetrahedral Fe^{3+} ions reduce this to 58 to be balanced by the ions in octahedral positions. Setting up our simultaneous equations, we know that if x is the number of Fe^{2+} ions and y the number of Fe^{3+}, then

$$x + y = 25 \qquad\qquad\qquad \textbf{40}$$

and adding up the charges, we get:

$$2x + 3y = 58 \qquad\qquad\qquad \textbf{41}$$

Substituting $x = 25 - y$ into equation 41, we find that $y = 8$, and so x must equal 17.

SAQ 34   Taking the titanium positions first: there are vacancies at all eight corners $(8 \times \frac{1}{8}) = 1$, and 1 vacancy in the centre of the cell, making 2 vacancies in all. There are no Ti^{2+} on cell edges. There are Ti^{2+} on only two of the faces—the top and bottom, which have 4 each—contributing $(2 \times 4 \times \frac{1}{2}) = 4$. There are 4 Ti^{2+} enclosed within the cell. Four edges have O^{2-} ions giving $(4 \times 1 \times \frac{1}{4}) = 1$; only the top and bottom faces have O^{2-} ions, and these have 5 each contributing $(2 \times 5 \times \frac{1}{2}) = 5$; there are 4$O^{2-}$ ions enclosed within the cell, making 10 in all. The overall content of this unit cell is thus Ti_8O_{10}, which corresponds to $TiO_{1.25}$.

SAQ 35   The effects of sharing are shown in Figure 105.

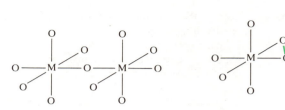

(a) sharing a corner M_2O_{11} (b) sharing an edge M_2O_{10}

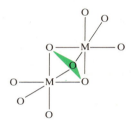

(c) sharing a face M_2O_9

Figure 105

SAQ 36   Figure 106 shows the shear structure with a unit cell added. Within the boundary, there is one group of four edge-sharing octahedra and four $[WO_6]$ octahedra. The formula is thus $W_4O_{11} + 4WO_3 = W_8O_{23}$.

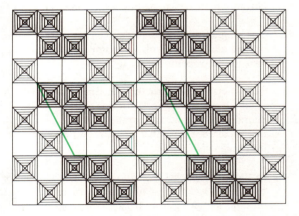

Figure 106 The structure of W_8O_{23} with a unit cell marked in green.

SAQ 37   Non-stoichiometric ZnO is an n-type semiconductor. Gallium ions entering the structure of ZnO have a charge of $+3$. If the Ga^{3+} substitutes for Zn^{2+} and the crystal maintains its stoichiometry, oxygen will be lost during the reaction. The electrons made available from the oxide ions becoming oxygen molecules will remain in the structure to effect the necessary charge compensation, thus enhancing the n-type semiconduction. An equation for the reaction is:

$$xGa_2O_3 + 2(1 - x)ZnO = 2Ga_xZn_{1-x}O + \tfrac{1}{2}xO_2 \qquad \textbf{42}$$

where for simplicity we have used 1 : 1 ZnO.

SAQ 38   A nickel atom has configuration $[Ar]4s^23d^8$. There are thus 10 electrons per atom, that is $10N$ electrons to feed into the 4s–4p and 3d bands. $0.6N$ of these are unpaired. The remainder $9.4N$ must be paired. You can see, then, that even in the ferromagnetic metals, most of the electrons are paired.

SAQ 39　With the lanthanides, the 4f orbitals are filled. These are likely to form narrow bands, and because there are seven 4f orbitals per atom, the density of states is likely to be even higher than for a 3d band. Ferromagnetism would be expected towards the end of the series, when the 4f band is more than half full if the situation is analogous to that for the first transition series.

Comment　In fact, the 4f band is even narrower than the 3d, and the 4f electrons are better thought of as localised. Ferromagnetism arises via an exchange mechanism of the sort to be described in the next Section, with the 5d and 6s bands playing the role that the oxide ions play. The metals that become ferromagnetic at reasonable temperatures are gadolinium ($T_C = 289$ K) and terbium ($T_C = 221$ K). These do occur towards the end of the series.

SAQ 40　Because the Mn atoms are further apart, the overlap of the 3d orbitals will be less. The 3d band will therefore be narrower than in manganese metal. With a narrower band, there is a smaller net promotion energy and larger interelectronic repulsion, and a state with a number of unpaired spins comparable to the number of atoms becomes favourable, and therefore the alloy would be ferromagnetic.

SAQ 41　A unit cell of the magnetic lattice is marked in green on Figure 107. The unit cell is now primitive (P) because the spin on the Cr atom at the body-centre is different from those at the corners, and so this Cr atom is no longer a lattice point. The cell dimension is *a*.

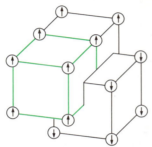

Figure 107

Comment　The antiferromagnetism of chromium is not as straightforward as it appears here. A complete explanation of its origin is beyond the scope of this Course, but you should note that the magnetic moments on each chromium atom correspond to only 0.6 of an unpaired electron. The electrons are thus largely delocalised and not localised on the chromium nuclei as you might have inferred from Figures 77 and 107

SAQ 42　The magnetically ordered unit cell is identical to the high-temperature (paramagnetic) unit cell, so all the layers of europium ions must be aligned with their spins parallel, giving a ferromagnetic compound.

SAQ 43　The inverse spinel structure for $ZnFe_2O_4$ means that the zinc(II) and iron(III) ions are on octahedral sites aligned with one another, while the remaining iron(III) ions are on tetrahedral sites aligned antiparallel. The net moment of the iron(III) ions is zero. As Zn^{2+} has no unpaired spins, there is no overall magnetic moment and the compound is antiferromagnetic.

SAQ 44　In the analogy, each person represents a Cooper pair and the linking of arms denotes the overlap between the pairs, giving an ordered system. Because of the ordered collective motion, scattering from defects—tripping over potholes—cannot take place! The analogy breaks down in that it allows for overlap only between adjacent pairs rather than over large numbers; also, the loss of superconductivity leads to the scattering of individual *electrons*, not of individual Cooper pairs.

SAQ 45　(a) The packing diagram for the A-type unit cell is shown in Figure 108. The contents of the unit cell are as follows: twelve oxygen atoms midway on cell edges, each shared by four unit cells: one calcium atom at the body-centre, eight titanium atoms at the cube corners—each shared by eight unit cells: $[(12 \times \frac{1}{4})O + 1Ca + (8 \times \frac{1}{8})Ti] = 3O + 1Ca + 1Ti = 1$ formula unit. The Ca in the centre is surrounded by twelve O atoms, and eight Ti atoms at the corners of a cube. A Ti atom is surrounded octahedrally by six O atoms, and by eight Ca atoms at the corners of a cube. An oxygen atom is linearly coordinated by two Ti atoms and by four Ca atoms in a square-planar configuration.

(b) Redrawing the structure now puts the B atom at the centre of the cell and the A atoms at the corners. The new cell has six face-centred oxygen atoms (Figure 109), and is called the B-type cell. Repeating either A-type or B-type cells gives the identical perovskite structure.

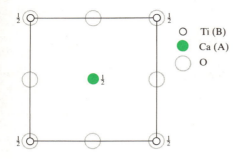

Figure 108 Packing diagram for a unit cell of perovskite, $ABO_3/CaTiO_3$.

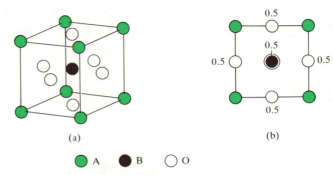

Figure 109 (a) B-type cell for the ABO_3 perovskite structure and (b) its packing diagram.

SAQ 46 The central section is a B-type perovskite unit cell where A = K/La, B = Ni/Cu, and O = F/O. K/La is now surrounded by *nine* oxygen atoms, whereas the A atom is surrounded by twelve oxygens in perovskite. The packing diagrams for the layers at $\frac{1}{6}$ and $\frac{1}{3}$ (and $\frac{5}{6}$ and $\frac{2}{3}$) in *c* are shown in Figure 110.

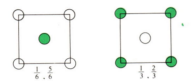

Figure 110 Layer sequences for K_2NiF_4.

ACKNOWLEDGEMENTS

Grateful acknowledgement is made to the following sources for material used in this Block.

Figure 36: R. Kirk and P. L. Pratt *Proceedings of the British Ceramic Society* (1967), vol. 9, Institute of Ceramics; *Figure 44:* R. Collongues *et al. Annual Review of Materials Science* (1979), vol. 9, copyright © 1979 Annual Reviews, Inc.; *Figure 50:* H. Rickert *Angewandte Chemie, International Edition* (English) (1978), vol. 17, copyright © 1978 VCH Verlagsgesellschaft; *Figure 51(a):* A. R. West *Solid State Chemistry and Its Applications* (1984), copyright © 1984 John Wiley & Sons, Ltd.; *Figure 54:* copyright © 1985 Electric Power Research Institute; *Figure 71(a):* R. Eisberg and R. Resnick *Quantum Physics of Atoms, Molecules, Solids, Nuclei and Particles*, copyright © 1974, 1985, John Wiley & Sons, Inc. (courtesy of H. J. Williams, Bell Telephone Laboratories); *Figure 71(b):* W. J. Moore *Seven Solid States* (1967), W. A. Benjamin, Inc., copyright © W. J. Moore; *Figure 90: Scientific American*, vol. 258, no. 2, February 1988, copyright © James Kilkelly, N.Y.; *Figure 91:* D. Bloor 'Organic conductors', *Chemistry in Britain*, vol. 19, no. 9, September 1983, copyright © 1983 The Royal Society of Chemistry; *Figure 92:* D. Bloor 'Plastics that conduct electricity', *New Scientist*, vol. 93, no. 1295, copyright © 1982 IPC Magazines; *Figure 96:* Courtesty of AT&T Archives.

S343 Inorganic Chemistry

Block 1   Introducing the transition elements

Block 2   Theory of metal-ligand interaction

Block 3   Transition-metal chemistry: the stabilities of oxidation states

Block 4   Structure, geometry and synthesis of transition-metal complexes

Block 5   Nuclear magnetic resonance spectroscopy

Block 6   Organometallic chemistry

Block 7   Bioinorganic chemistry

Block 8   Solid-state chemistry

Block 9   Actinide chemistry and the nuclear fuel cycle